PHY-122

ELECTRICITY AND MAGNETISM

[2 Credits]

Physics : Paper-II

For

First Year B.Sc. Semester-II

As Per New Syllabus of CBCS Pattern

From June - 2019

Dr. P. S. TAMBADE
M. Sc., Ph.D.
Department of Physics,
Prof. Ramkrishna More, A.C.S. College,
Akurdi, Pune 411044.

B. M. LAWARE
M. Sc., M.Phil.
Head, Department of Physics,
Prof. Ramkrishna More, A.C.S. College,
Akurdi, Pune 411044

Dr. S. D. AGHAV
M. Sc., M.Phil., Ph.D.
Ex. Vice Principal,
Baburaoji Gholap College,
Sangavi, Pune 411027.

V. K. DHAS
M. Sc., M. Phil
Ex. Head, Department of Physics,
New Arts, Commerce & Science College,
Ahmednagar.

N5027

F.Y.B.Sc. Physics : Electricity and Magnetism (PHY-122) (P-II)
First Edition :   November 2019   ISBN 978-93-89533-64-4
©   :   Authors

Published By:
NIRALI PRAKASHAN
Abhyudaya Pragati, 1312, Shivaji Nagar
Off J.M. Road, PUNE – 411005
Tel - (020) 25512336/37/39, Fax - (020) 25511379
Email : niralipune@pragationline.com

➢ DISTRIBUTION CENTRES

PUNE

Nirali Prakashan : 119, Budhwar Peth, Jogeshwari Mandir Lane, Pune 411002,
(For orders within Pune)   Maharashtra, Tel : (020) 2445 2044, Mobile : 9657703145
Email : niralilocal@pragationline.com

Nirali Prakashan : S. No. 28/27, Dhayari, Near Asian College Pune 411041
(For orders outside Pune)   Tel : (020) 24690204; Mobile : 9657703143
Email : bookorder@pragationline.com

MUMBAI

Nirali Prakashan : 385, S.V.P. Road, Rasdhara Co-op. Hsg. Society Ltd.,
Girgaum, Mumbai 400004, Maharashtra;
Mobile : 9320129587 Tel : (022) 2385 6339 / 2386 9976,
Fax : (022) 2386 9976
Email : niralimumbai@pragationline.com

➢ DISTRIBUTION BRANCHES

JALGAON

Nirali Prakashan : 34, V. V. Golani Market, Navi Peth, Jalgaon 425001,
Maharashtra, Tel : (0257) 222 0395, Mob : 94234 91860;
Email : niralijalgaon@pragationline.com

KOLHAPUR

Nirali Prakashan : New Mahadvar Road, Kedar Plaza, 1st Floor Opp. IDBI Bank,
Kolhapur 416 012, Maharashtra. Mob : 9850046155;
Email : niralikolhapur@pragationline.com

NAGPUR

Nirali Prakashan : Above Maratha Mandir, Shop No. 3, First Floor,
Rani Jhanshi Square, Sitabuldi, Nagpur 440012, Maharashtra
Tel : (0712) 254 7129;
Email : niralinagpur@pragationline.com

DELHI

Nirali Prakashan : 4593/15, Basement, Agarwal Lane, Ansari Road, Daryaganj
Near Times of India Building, New Delhi 110002
Mob : 08505972553, Email : niralidelhi@pragationline.com

BENGALURU

Nirali Prakashan : Maitri Ground Floor, Jaya Apartments, No. 99, 6th Cross,
6th Main, Malleswaram, Bengaluru 560003, Karnataka;
Mob : 9449043034
Email: niralibangalore@pragationline.com
Other Branches : Hyderabad, Chennai

niralipune@pragationline.com | www.pragationline.com

Also find us on   www.facebook.com/niralibooks

Preface ...

This book entitled **"Electricity and Magnetism"** is designed for the students preparing for F.Y.B.Sc. Semester-II examination of Savitribai Phule Pune University. It is a purpose of textbook to give systematic exposition of CBCS pattern syllabus implemented from June 2019.

Beginning with its fundamentals the subject has been developed systematically and logically with the emphasis on the physical explanation supported adequately by mathematical formulation. The clarity and systematic representation will make the book intelligible to a beginner. Solved Problems have been included to illustrate applications of theoretical principles.

The book entitled, "Electricity and Magnetism" is written in a simple and lucid language as possible. It will be helpful for understanding of Electrostatics, Dielectrics, Magnetization, Magnetostatics and Magnetic Properties of Materials. Multiple choice questions and true and false type questions are also added which are useful for competitive exam. The content included in the section will surely increase problem solving skills.

Authors have done sincere efforts to explain above topics in a simple and lucid language as possible.

Although the book is not voluminous, every useful and important information has been added. Authors sincerely feel that the work will adequately meet the needs of F.Y.B.Sc. students for a good and concise book which will stimulate a genuine interest among students in the subject.

Authors sincerely thank Shri. Dineshbhai Furia, Mr. Jignesh Furia, and entire staff of Nirali Prakashan especially Mr. Santosh Bare, Mr. Kiran Velankar and Mrs. Anjali Muley for their constant encouragement in this endeavour.

Sufficient care has been taken to avoid misprints, checking solutions and answers of numerical problems. However, authors will most gratefully accept suggestions for improving the book and making it more useful.

AUTHORS

Syllabus ...

1.	**ELECTROSTATICS**	**(08 Lectures)**

1.1 Revision of Coulomb's Law

    1.1.1 Statement

    1.1.2 Variation of Forces with Distances

1.2 Superposition Principle

    1.2.1 Statement

    1.2.2 Explanation with Illustration

1.3 Energy of System of Charges

1.4 Concept of Electric Field

    1.4.1 Due to Point Charge

    1.4.2 Due to Group Charges

1.5 Concept of Electric Flux

1.6 Gauss's Law in Electrostatics

- Problems

2.	**DIELECTRICS**	**(08 Lectures)**

2.1 Introduction to Dielectric Materials

2.2 Electric Dipole

    2.2.1 Electric Dipole

    2.2.2 Dipole Moment

2.3 Electric Potential and Intensity at any Point due to Dipole

2.4 Torque on a Dipole placed in an Electric Field

2.5 Polar and Non-polar Molecules

2.6 Electric Polarization of Dielectric Material

2.7 Gauss's Law of Dielectrics

2.8 Electric Vectors and its Relation

- Problems

3.	**MAGNETIZATION**	**(07 Lectures)**

3.1 Introduction to Magnetization

3.2 Magnetic Materials

3.3 Types of Magnetic Materials

    3.3.1 Diamagnetic Materials

    3.3.2 Paramagnetic Materials

    3.3.3 Ferromagnetic Materials

    3.3.4 Antiferromagnetic Materials

3.4 Bohr Magneton

- Problems

4. MAGNETOSTATICS (07 Lectures)

4.1 Introduction to Magnetization
4.2 Magnetic Induction and Intensity of Magnetization
4.3 Biot-Savart's Law
    4.3.1 Statement
    4.3.2 Long Straight Conductor
    4.3.3 Circular Coil
4.4 Ampere's Circuital Law
    4.4.1 Statement
    4.4.2 Field of Solenoid
    4.4.3 Field of Toroid
4.5 Gauss's Law for Magnetism
• Problems

5. MAGNETIC PROPERTIES OF MATERIALS (06 Lectures)

5.1 Definitions
    5.1.1 Magnetization ($\vec{M}$)
    5.1.2 Magnetic Intensity ($\vec{H}$)
    5.1.3 Magnetic Induction ($\vec{B}$)
    5.1.4 Magnetic Susceptibility (χ_m)
    5.1.5 Magnetic Permeability (μ)
5.2 Relation between B, M and H
5.3 Hysteresis and Hysteresis Curve
5.4 Ferrite Materials and its Applications
• Problems

Reference Books :

1. Fundamentals of Physics : Halliday Resnik and Walkar, 8th Edition.
2. Electromagnetics : B. B. Laud.
3. Foundations of Electromagnetic Theory : Reitz, Milford, Christey.
4. Electricity and Electronics : D.C. Tayal, Himalaya Publishing House, Mumbai.
5. Introduction to Electrodynamics : D. G. Griffith.
6. Electricity and Magnetism : BrijLal, Subramanyan, Ratan Prakashan (Revised Edition, 1997).
7. Electricity and Magnetism : Khare, Shrivastav (Revised Edition, 1997).

❑❑❑

Contents ...

□□□

PATTERN OF UNIVERSITY EXAM QUESTION PAPER

Total Marks : 35   **Duration : 2 Hours**

Note :
- (a) Q.1 is compulsory.
- (b) Solve any three questions from Q.2 to Q.5.
- (c) Questions 2 to 5 carry equal marks.

1. Solve any five of the following :
   - (a)
   - (b)
   - (c)
   - (d)
   - (e)
   - (f)

   (Ask four tricky questions and two questions based on problem solving, if applicable.)   (5)

2. (A) Describe – type of question(s) with internal options.   (6)

   (B) Short question, but tricky.   (4)

3. (A) Explain – type of question(s) with internal options.   (6)

   (B) Problem based question, if applicable otherwise justification type – question.   (4)

4. (A) Discuss - type of question(s) with internal options.   (6)

   (B) Problem based question, if applicable otherwise tricky and signified question.   (4)

5. Write short notes on any four of the following :   (10)
   - (A) Principle based.
   - (B) System based.
   - (C) Structure based.
   - (D) Descriptive based.
   - (E) Working based.
   - (F) Model based.

❑❑❑

Chapter **1**...

Electrostatics

Contents ...

If you place a plastic comb near tiny scraps of paper, nothing happens. But immediately you comb your hair and or strike the comb with fur, it will attract the paper scraps. This phenomenon called "static cling" occurs between many different objects.

The photograph herewith shows small pieces of paper attracted by charged CD.

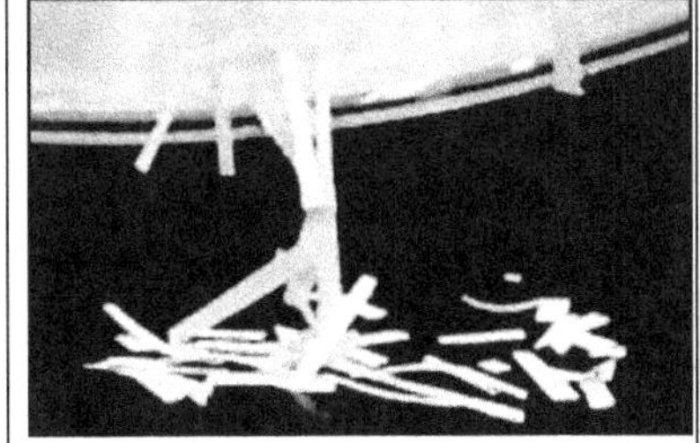

Charles-Augustin de **Coulomb**, (born June 14, 1736, Angoulême, France - died August 23, 1806, Paris), French physicist best known for the formulation of **Coulomb's** law, which states that the force between two electrical charges is proportional to the product of the charges and inversely proportional to the square of the distance between them and is along the line joining them.

Charles-Augustin de Coulomb

Introduction

- In the ancient days about 600 B.C., the great philosopher Thales of Greece discovered that when a fossil material called amber rubbed with wool, it attracts light objects like dust, small pieces of paper, etc. At the end of 16th century A.D., William Gilbert found that this phenomenon is not restricted to amber and wool, but occurs when the substances like glass, ebonite, etc. are rubbed with suitable substances. He proposed that a body after rubbing becomes charged or electrified and acquires the property of attracting light objects. As this phenomenon was first observed with amber, the word electricity is derived from *elektron* which is Greek name of amber.

- It is further observed that two glass rods rubbed with silk repel each other or two ebonite rods rubbed with fur repel each other, but an ebonite rod rubbed with fur attracts a glass rod rubbed with silk. These observations show that glass rod (rubbed with silk) and ebonite rod (rubbed with fur) have different kinds of charges. Benjamin Franklin, an American physicist named the kind of electric charge on the glass rod as positive charge and kind of electric charge on the ebonite rod as negative charge. From the above observations, it is concluded that, *like charges repel each other and unlike charges attract each other.*

- The electrification of bodies by friction can be explained as follows. In an atom of any substance, the total negative charge carried by all electrons is equal to positive charge (due to protons) of the nucleus. Therefore, an atom is electrically neutral particle. Now when two bodies are rubbed together, there is a transfer of electrons from one body to other body. As a result, the body which gains electron becomes negatively charged, while the body which loses electron acquires an equal amount of positive charge. Thus whenever two bodies are rubbed with each other, they acquire opposite charges of equal magnitude and behave accordingly.

- *The study of electricity or electric charges at rest is known as electrostatics.*
- *It is known that charge is a scalar quantity and S.I. unit of charge is coulomb.* On the basis of atomic theory, it was proved that charge carried by an object is not continuous, but it is an integral multiple of certain smallest unit of charge. The smallest unit of charge is the charge on an electron. (The magnitude of charge on electron is 1.6×10^{-19} coulomb). So charge carried by any body is always equal to $\pm ne$, where $n = 0, 1, 2, ...,$ etc. This means that an object may have a charge $\pm e, \pm 2e, ...$ etc. but never like a charge of $\pm 1.5\ e, \pm 1.7\ e$ etc. Thus a charge is said to have a discrete nature or is said to be quantized. In the present chapter, we shall study two fundamental laws namely Coulomb's law and Gauss's law. These laws are based on experimental studies and they are independent. Based on Coulomb's law, the concept of electric field is introduced. Special problems are solved using Coulomb's law and Gauss's law.

1.1 Revision of Coulomb's Law

We have seen that like charges repel each other, while unlike charges attract each other. On the basis of experimental study, Charles Augustin Coulomb (1736-1806) formulated the fundamental law of electric force between two stationary charged particles which is known as Coulomb's law.

1.1.1 Statement of Coulomb's Law

This law states that *the force of attraction or repulsion between two point charges* * *at rest is directly proportional to the product of magnitude of the charges and inversely proportional to the square of the distance between them and is along the line joining them.* Thus, if two point charges q_1 and q_2 are separated by a distance r in some medium, then the magnitude of the force, which one exerts on the other, is given by,

$$F \propto q_1 q_2 \quad \text{and} \quad F \propto \frac{1}{r^2}$$

---

* *Charged bodies whose sizes are much smaller than the distance between them are referred as point charges.*

Combining above equations, we get

$$F \propto \frac{q_1 q_2}{r^2} \quad \text{or} \quad F = \beta \frac{q_1 q_2}{r^2}$$

where β is a constant of proportionality, which we call the Coulomb's constant.

The value of β depends upon the system of units we use and the nature of the medium in which two charges are situated.

In S.I. units, the unit of charge is coulomb (C), the unit of distance is meter (m), the unit of force is newton (N). If charges are situated in free space (air or vacuum), then in S.I. units, the constant β is equal to $\frac{1}{4\pi\varepsilon_o}$.

i.e. $$\beta = \frac{1}{4\pi\varepsilon_o}$$

where ε_o is called permittivity of free space. Thus, Coulomb's law in free space can be expressed in S.I. units as

$$F = \frac{1}{4\pi\varepsilon_o} \frac{q_1 q_2}{r^2} \qquad \text{... (1.1)}$$

The value of ε_o can be determined experimentally and it has been found to be $\varepsilon_o = 8.854 \times 10^{-12}$ C^2/Nm^2

Hence $$\frac{1}{4\pi\varepsilon_o} = 8.98 \times 10^9 \text{ N.m}^2/\text{C}^2 \approx 9 \times 10^9 \text{ N.m}^2/\text{C}^2$$

If the charges are situated in some other medium, then Coulomb's law (in S.I. units) can be expressed as

$$F = \frac{1}{4\pi\varepsilon} \frac{q_1 q_2}{r^2}$$

where ε is permittivity of the medium in which charges are situated.

The ratio of permittivity of the medium to the permittivity of free space is called the dielectric constant (k) of that medium.

i.e. $$k = \frac{\varepsilon}{\varepsilon_o} \quad \text{or} \quad \varepsilon = k\varepsilon_o \qquad \text{... (1.2)}$$

Therefore, in S.I. units, Coulomb's law for any medium can be written as $$F = \frac{1}{4\pi k\varepsilon_o} \frac{q_1 \, q_2}{r^2} \qquad \text{... (1.3)}$$

The dielectric constant (k), being a ratio of two similar quantities, is a pure number and has no dimensions and no units.

For air or vacuum, $k = 1$, while for any other medium, dielectric constant $k > 1$.

The equation (1.3) represents scalar form of Coulomb's law in S.I. units.

Vector Form of Coulomb's Law (April 10, Oct. 15)

To write Coulomb's law in vector form, let us use following notations :

$$\vec{F}_{12} = \text{Force acting on charge } q_1 \text{ due to } q_2$$

$$\vec{F}_{21} = \text{Force acting on charge } q_2 \text{ due to } q_1$$

$$\vec{r}_{12} = \text{Displacement measured from charge } q_1 \text{ to } q_2$$

$$\vec{r}_{21} = \text{Displacement measured from charge } q_2 \text{ to } q_1$$

$$\hat{r}_{12} = \text{Unit vector in the direction of } \vec{r}_{12}$$

$$\hat{r}_{21} = \text{Unit vector in the direction of } \vec{r}_{21}$$

$$r = \text{Distance between two charges, } r = |\vec{r}_{12}| = |\vec{r}_{21}|$$

(a) The electric force between two like charges : The electric force between two like charges is of repulsive nature (shown in Fig. 1.1).

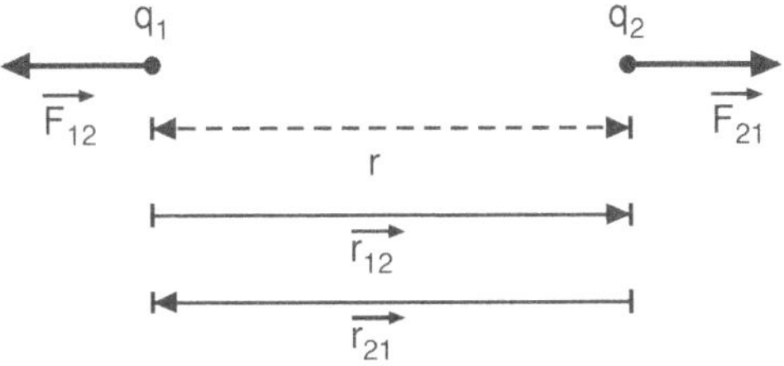

Fig. 1.1 : Electric force between two like charges

The magnitude of force acting on q_1 due to charge q_2 is

$$F_{12} = \frac{1}{4\pi\varepsilon_o} \frac{q_1\, q_2}{r^2}$$

In Fig. 1.1, $\vec{F}_{12}$ and $\vec{r}_{21}$ are in the same direction, so F_{12} in vector form may be written as

$$\vec{F}_{12} = \frac{1}{4\pi\varepsilon_o} \frac{q_1\, q_2}{r^2} \hat{r}_{21}$$

where $\hat{r}_{21}$ is a unit vector along $\vec{r}_{21}$.

But,
$$\hat{r}_{21} = \frac{\vec{r}_{21}}{|\vec{r}_{21}|} = \frac{\vec{r}_{21}}{r}$$

$\therefore$
$$\vec{F}_{12} = \frac{1}{4\pi\varepsilon_o}\frac{q_1\,q_2}{r^3}\,\vec{r}_{21} \qquad \dots (1.4)$$

Similarly, force acting on q_2 due to q_1 can be written as

$$\vec{F}_{21} = \frac{1}{4\pi\varepsilon_o}\frac{q_1\,q_2}{r^3}\,\vec{r}_{12} \qquad \dots (1.5)$$

From Fig. 1.1, $\vec{r}_{12} = -\vec{r}_{21}$

$\therefore$
$$\vec{F}_{12} = -\vec{F}_{21} \qquad \dots (1.6)$$

(b) The electric force between two unlike charges : The electric force between two unlike or opposite charges is of attractive nature (shown in Fig. 1.2).

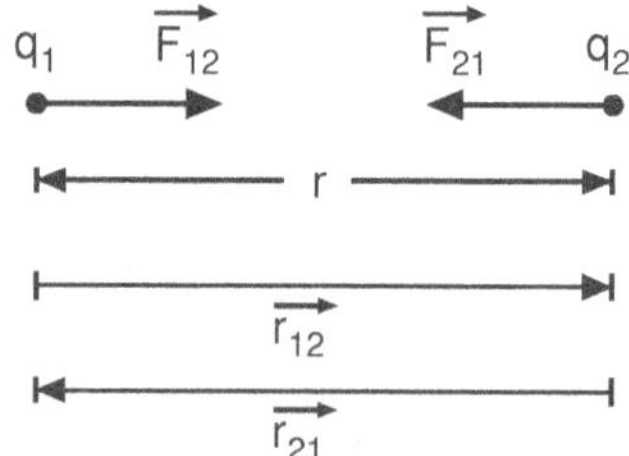

Fig. 1.2 : Electric force between two unlike charges

The magnitude of the force acting on charge q_1 due to charge q_2 is

$$F_{12} = \frac{1}{4\pi\varepsilon_o}\frac{q_1\,q_2}{r^2}$$

In Fig. 1.2, $\vec{F}_{12}$ and $\vec{r}_{12}$ are in the same direction, so $\vec{F}_{12}$ in vector form may be written as
$$\vec{F}_{12} = \frac{1}{4\pi\varepsilon_o}\frac{q_1\,q_2}{r^2}\,\hat{r}_{12}$$

But
$$\hat{r}_{12} = \frac{\vec{r}_{12}}{|\vec{r}_{12}|} = \frac{\vec{r}_{12}}{r}$$

$\therefore$
$$\vec{F}_{12} = \frac{1}{4\pi\varepsilon_o}\frac{q_1\,q_2}{r^3}\,\vec{r}_{12} \qquad \dots (1.7)$$

Similarly, the force acting on charge q_2 due to charge q_1 can be written as

$$\vec{F}_{21} = \frac{1}{4\pi\varepsilon_o}\frac{q_1\,q_2}{r^3}\vec{r}_{21} \qquad \text{... (1.8)}$$

But from Fig. 1.2, $\vec{r}_{12} = -\vec{r}_{21}$

$$\therefore \qquad \vec{F}_{12} = -\vec{F}_{21} \qquad \text{... (1.9)}$$

By observing equations (1.6) and (1.9), it can be concluded that whenever two charges are facing each other, these charges experience equal and opposite forces.

Following observations can be noted in connection with Coulomb's force between two electric charges :

(i)   Coulomb's law is strictly applicable in case of point charges at rest. While dealing with force between charges in motion, the expression has to be modified.

(ii) This law holds good over the wide range from 10^{-13} meter to many kilometers. For distances less than 10^{-13} meter, it breaks down due to predominance of nuclear forces.

(iii) This force is a central force i.e. a force acting along the line joining the two charges.

(iv) The force between two charges is not affected by the presence of other charges.

1.1.2 Variation of Forces with Distances

We have seen that, the electrostatic or electric force between two charges q_1 and q_2 separated by a distance r predicted by Coulomb's law is given by

$$F_c = \beta\,\frac{q_1 q_2}{r^2} \qquad \text{... (1.10)}$$

where β is constant ($\beta = 9.0 \times 10^9$ Nm^2/C).

The gravitational force between two particles of masses m_1 and m_2 separated by a distance r predicted by Newton's law is given by

$$F_g = G\,\frac{m_1 m_2}{r^2} \qquad \text{... (1.11)}$$

where G is constant ($G = 6.674 \times 10^{-11}$ m^3/kg s^2).

From above equations (1.10) and (1.11), it is clear that both forces are inversely proportional to the square of the distance of separation, so both are called inverse-square laws.

If we double the distance between particles, then the force between them (gravitational or electrostatic) decreases by one-forth. Fig. 1.3 shows how the force decreases or falls off as a function of separation between two particles.

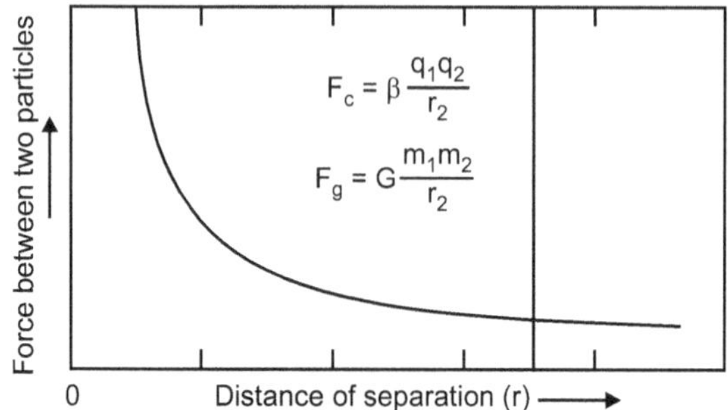

Fig. 1.3 : Variation of forces with distances

1.2 Superposition Principle

- Coulomb's law gives the electrostatic force on one charge due to another charge. If there are more than two charged particles or group of charges, then force on any charge due to other charges can be obtained by using the principle of superposition.

- To understand the use of principle of superposition to calculate the total force on any charge due to other charges, consider simple example of an assembly of three charges q_1, q_2 and q_3 situated at different points in a given region. As the force acting on any charge due to another charge is not affected by the presence of other charges. So using Coulomb's law, we can calculate the force separately for each pair of charge.

- For example, force on charge q_1 due to q_2 is $\vec{F}_{12}$ and force on charge q_1 due to q_3 is $\vec{F}_{13}$.

- The vector sum of $\vec{F}_{12}$ and $\vec{F}_{13}$ gives the total force on charge q_1. Thus, for a system consisting of three charges, the total force on charge q_1 is $\vec{F}_1 = \vec{F}_{12} + \vec{F}_{13}$. For an assembly of charges consisting of charges q_1, q_2, ..., q_n, situated at different points in a given region, the total force on any charge such as q_1 due to other charges is

$$\vec{F}_1 = \vec{F}_{12} + \vec{F}_{13} + \vec{F}_{14} + \ldots\ldots + \vec{F}_{1n}$$

- The above equation is mathematical representation of principle of superposition.

1.2.1 Statement

The superposition principle *states that all the charges when placed near each other, behave independent of each other and the net force on any particular charge is the vector sum of the forces acting on that charge due to each of the other charge.*

1.2.2 Explanation with Illustration

The following example illustrates the superposition principle.

Consider a group of three charges q_1, q_2 and q_3 situated at points having position vectors $\vec{r_1}$, $\vec{r_2}$ and $\vec{r_3}$ respectively as shown in Fig. 1.4.

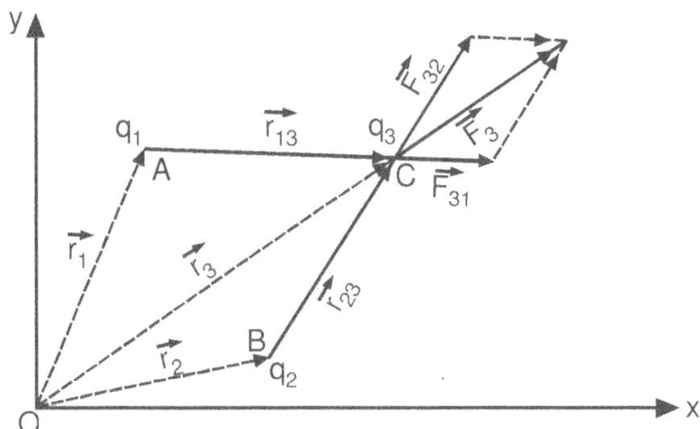

Fig. 1.4 : Principle of superposition

In Fig. 1.4, $\vec{r}_{13}$ = Displacement vector from charge q_1 to q_3

$\vec{r}_{23}$ = Displacement vector from charge q_2 to q_3

Let us calculate the total force on charge q_3 due to q_1 and q_2 by using the principle of superposition.

(a) In Fig. 1.4, $\vec{F}_{31}$ and $\vec{r}_{13}$ are in same direction.

According to Coulomb's law, force on charge q_3 due to q_1 is

$$\vec{F}_{31} = \frac{1}{4\pi\varepsilon_o} \frac{q_1\, q_3}{|\vec{r}_{13}|^3} \vec{r}_{13}$$

(b) Force on charge q_3 due to q_2 is

$$\vec{F}_{32} = \frac{1}{4\pi\varepsilon_o} \frac{q_2\, q_3}{|\vec{r}_{23}|^3} \vec{r}_{23}$$

Net force on charge q_3 due to q_1 and q_2 is

$$\vec{F_3} = \vec{F_{31}} + \vec{F_{32}} = \frac{1}{4\pi\varepsilon_o}\left[\frac{q_1 q_3}{|\vec{r_{13}}|^3}\vec{r_{13}} + \frac{q_2 q_3}{|\vec{r_{23}}|^3}\vec{r_{23}}\right] \qquad \ldots (1.12)$$

The direction of resultant force $\vec{F_3}$ is as shown in Fig. 1.4.

Similarly, if we have a group of n charges $q_1, q_2, q_3, \ldots, q_n$ situated at points having position vectors $\vec{r_1}, \vec{r_2}, \vec{r_3}, \ldots, \vec{r_n}$ respectively, then force on charge q_i due to all other charges can be written as

$$\vec{F_i} = \frac{1}{4\pi\varepsilon_o}\left[\frac{q_1 q_i}{|\vec{r_{1i}}|^3}\vec{r_{1i}} + \frac{q_2 q_i}{|\vec{r_{2i}}|^3}\vec{r_{2i}} + \ldots\ldots \frac{q_n q_i}{|\vec{r_{ni}}|^3}\vec{r_{ni}}\right]$$

or,

$$\vec{F_i} = \frac{1}{4\pi\varepsilon_o}\sum_{\substack{j=1 \\ j \neq i}}^{n}\frac{q_j q_i}{|\vec{r_{ji}}|^3}\vec{r_{ji}}$$

i.e.

$$\vec{F_i} = \sum_{\substack{j=1 \\ j \neq i}}^{n}\vec{F_{ij}} \qquad \ldots (1.13)$$

It should be noted that this principle is applicable when charges are located in free space (i.e. not inside, material medium) and are separated by a distance involving classical length scales. Inside certain material medium, non-linear effects may encounter. So to use principle of superposition, its validity must be tested experimentally.

Solved Problems

Problem 1.1 : *Calculate and compare the electrostatic force and the gravitational force between the proton and electron by a hydrogen atom in its ground state, in which the proton-electron separation is 5.3×10^{-11} m.*

Solution : The electrostatic or Coulomb's force is given by

$$F_c = \beta\frac{q_1 q_2}{r^2} = \frac{9 \times 10^9 \times 1.6 \times 10^{-19} \times 1.6 \times 10^{-19}}{(5.3 \times 10^{-11})^2}$$

$$F_c = 8.21 \times 10^{-8} \text{ N (The force is attractive)}$$

The gravitational force is given by

$$F_g = G\frac{m_1 m_2}{r^2}$$

$$= \frac{6.674 \times 10^{-11} \times 9.11 \times 10^{-31} \times 1.67 \times 10^{-27}}{(5.3 \times 10^{-11})^2}$$

$$F_g = 3.61 \times 10^{-47} \text{ N (The force is attractive)}$$

Now, to compare, let us divide the larger force by a smaller force.

$$\therefore \qquad \frac{F_c}{F_g} = \frac{8.21 \times 10^{-8}\ N}{3.61 \times 10^{-47}\ N} \approx \mathbf{2.3 \times 10^{39}} \qquad \textbf{... Ans.}$$

Thus, the electrostatic force is about 40 orders of magnitude stronger than gravitational force.

Problem 1.2 : *Three point charges $q_1 = 1.2\ \mu C$, $q_2 = -0.60\ \mu C$ and $q_3 = 0.20\ \mu C$ are located at the corners of a right angled triangle as shown in Fig. 1.5. What is the electric force on a charge $0.20\ \mu C$ due to the other two charges ? The distances of separation between the charges are shown in Fig. 1.5.*

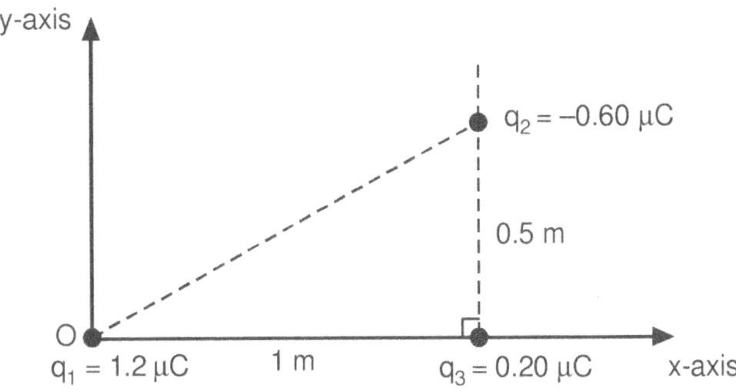

Fig. 1.5 : Location of charges q_1, q_2, q_3 at three corners of a triangle

Solution : To find the total force on charge q_3 due to other charges q_1 and q_2, we have to find the force on q_3 due to charge q_1 and due to q_2 separately.

(i)   The charges q_1 and q_3 are both positive. The force $\overrightarrow{F_{31}}$ on charge q_3 due to q_1 is repulsive and it can be shown by a vector pointing in the positive X-direction as shown in Fig. 1.5.

The magnitude of force $\overrightarrow{F_{31}}$ on charge $q_3 = 0.20\ \mu C$ due to charge $q_1 = 1.2\ \mu C$ is

$$F_{31} = \frac{1}{4\pi\varepsilon_o} \frac{|q_1|\,|q_3|}{r_{13}^2}$$

$$= 8.99 \times 10^9 \times \frac{1.2 \times 10^{-6} \times 0.2 \times 10^{-6}}{(1)^2}$$

$$F_{31} = 2.16 \times 10^{-3}\ N \qquad \text{... (1)}$$

(ii) The charge q_2 is negative, so it attracts charge q_3 along the line joining the two charges. The force $\overrightarrow{F_{32}}$ points in the positive Y-direction as shown in Fig. 1.6.

The magnitude of force $\overrightarrow{F_{32}}$ on charge q_3 = 0.20 μC due to q_2 = –0.60 μC is

$$F_{32} = \frac{1}{4\pi\varepsilon_o}\frac{|q_2||q_3|}{r_{23}^2}$$

$$= 8.99 \times 10^9 \times \frac{0.60 \times 10^{-6} \times 0.20 \times 10^{-6}}{(0.5)^2}$$

$$= 4.32 \times 10^{-3} \text{ N} \qquad\qquad ... (2)$$

From equations (1) and (2), we get

$$F_{32} > F_{31}$$

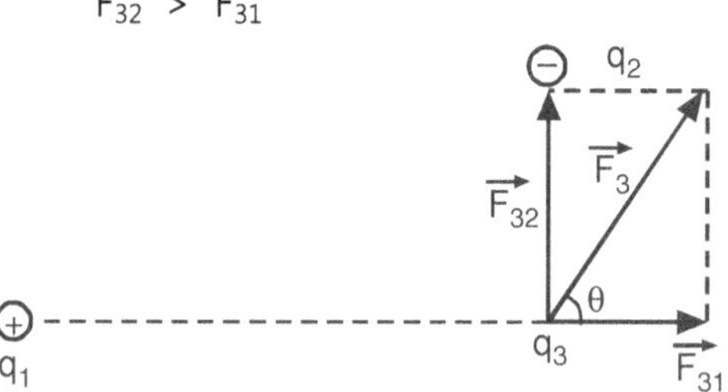

Fig. 1.6 : The direction of forces $\overrightarrow{F_{31}}$, $\overrightarrow{F_{32}}$ and $\overrightarrow{F_3}$

The addition of two vectors $\overrightarrow{F_{31}}$ and $\overrightarrow{F_{32}}$ gives the total force $\overrightarrow{F_3}$. Since two vectors are perpendicular to each other, the magnitude of F_3 is

$$F_3 = \sqrt{F_{31}^2 + F_{32}^2} = \sqrt{(2.16 \times 10^{-3})^2 + (4.32 \times 10^{-3})^2}$$

$$F_3 = 4.83 \times 10^{-3} \text{ N}$$

From Fig. 1.6, the direction of the force $\overrightarrow{F_3}$ is

$$\tan\theta = \frac{F_{32}}{F_{31}} = \frac{4.32 \times 10^{-3}}{2.16 \times 10^{-3}} = 2$$

$$\therefore \qquad\qquad \theta = \mathbf{63.43°} \qquad\qquad ... \textbf{Ans.}$$

Thus, the resultant force makes an angle 63.43° with X-axis.

Problem 1.3 : *Four point charges 10 μC, 15 μC, 10 μC and –20 μC are placed on the four corners A, B, C and D respectively of a square ABCD of side 4 m. Calculate the total force on a charge 15 μC due to other three charges.*

Solution : To find the total force on a charge 15 µC due to other charges 10 µC, 10 µC and −20 µC, we have to find the force on 15 µC due to each of the charge separately.

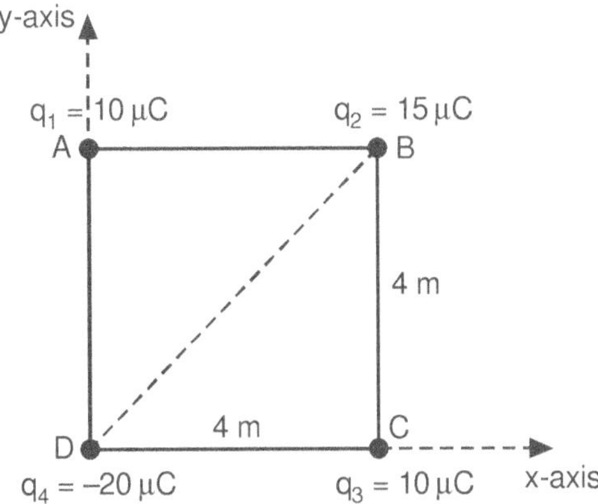

Fig. 1.7 : Location of point charges on the corners A, B, C and D of a square

(i) The charges q_1 and q_2 are both positive, so the force $\overrightarrow{F_{21}}$ on a charge q_2 due to q_1 is repulsive. The direction of the force $\overrightarrow{F_{21}}$ is as shown in Fig. 1.8.

The magnitude of force $\overrightarrow{F_{21}}$ on charge q_2 due to q_1 is

$$F_{21} = \frac{1}{4\pi\varepsilon_o} \frac{|q_2|\,|q_1|}{r_{12}^2}$$

But,　　$|q_2| = 15 \times 10^{-6}, \quad |q_1| = 10 \times 10^{-6}\ C$

and　　$r_{12} = r = 4\ m$

$\therefore$　　$F_{21} = \dfrac{8.99 \times 10^9 \times 15 \times 10^{-6} \times 10 \times 10^{-6}}{(4)^2}$

$F_{21} = 8.42 \times 10^{-2}\ N$

(ii) The charges q_3 and q_2 are both positive, so the force $\overrightarrow{F_{23}}$ on charge q_2 due to q_3 is repulsive. The magnitude of force $\overrightarrow{F_{23}}$ on charge q_2 due to charge q_3 is　　$F_{23} = \dfrac{1}{4\pi\varepsilon_o} \dfrac{|q_2|\,|q_3|}{r^2}$

$$F_{23} = \frac{8.99 \times 10^9 \times 15 \times 10^{-6} \times 10 \times 10^{-6}}{(4)^2}$$

$F_{23} = 8.42 \times 10^{-2}\ N$

The direction of $\overrightarrow{F_{23}}$ is as shown in Fig. 1.8.

(iii) The magnitude of force $\vec{F}_{24}$ on charge q_2 due to charge q_4 is

$$F_{24} \;=\; \frac{1}{4\pi\varepsilon_o}\frac{|q_2|\,|q_4|}{r^2}$$

The charges q_2 and q_4 are of opposite nature, so the force $\vec{F}_{24}$ is attractive and its direction is as shown in Fig. 1.8. The magnitude of force $\vec{F}_{24}$ is

$$F_{24} = \frac{8.99 \times 10^{9} \times 15 \times 10^{-6} \times 20 \times 10^{-6}}{(4\sqrt{2})^2} = 8.42 \times 10^{-2}\ \text{N}$$

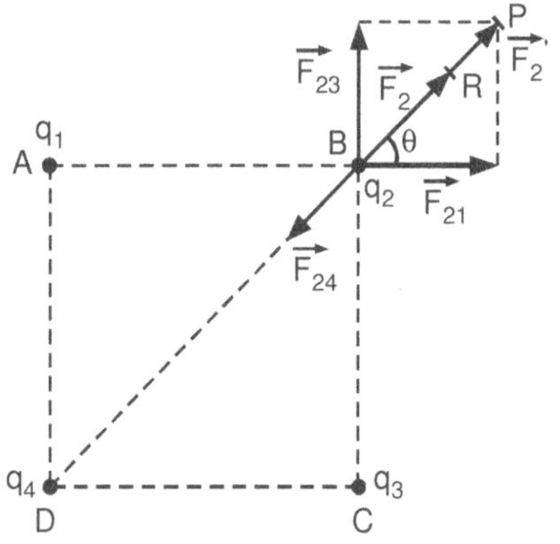

Fig. 1.8 : The directions of forces $\vec{F}_{21}$, $\vec{F}_{23}$, $\vec{F}_{24}$ and resultant force $\vec{F}_2$

Since two vectors $\vec{F}_{21}$ and $\vec{F}_{23}$ are perpendicular to each other, the magnitude of resultant force on charge q_2 due to q_1 and q_3 is

$$F_2' \;=\; \sqrt{F_{21}^2 + F_{23}^2} = \sqrt{(8.42 \times 10^{-2})^2 + (8.42 \times 10^{-2})^2}$$

$$F_2' \;=\; 11.9 \times 10^{-2}\ \text{N}$$

From Fig. 1.8, the direction of $\vec{F}_2'$ ($\vec{F}_2' = \vec{BP}$)

$$\tan \theta' \;=\; \frac{F_{23}}{F_{21}} = 1$$

$$\theta' \;=\; 45°$$

Thus, the resultant force on charge q_2 due to q_1 and q_3 makes an angle 45° with the X-axis.

Since forces $\vec{F}_{24}$ and $\vec{F'}_2$ are oppositely directed, the magnitude of resultant force $\vec{F}_2$ $(\vec{F}_2 = \vec{BR})$ on charge q_2 due to all other charges is

$$F_2 \ = \ F'_2 - F_{24} = 11.9 \times 10^{-2} - 8.42 \times 10^{-2} = 3.48 \times 10^{-2} \ N$$

Thus, the magnitude of resultant force F_2 on charge q_2 due to all other charges is 3.48×10^{-2} N.

The resultant force makes an angle of 45° with the X-axis as shown in Fig. 1.8.

1.3 Energy of System of Charges

Before we study the electric potential energy of a system of charges, let us first study electric potential and calculate electric potential of the point charge +q at a point situated at a distance r from it.

1.3.1 Electric Potential due to a Point Charge

The electric potential at any point in an electric field is defined as the *work done in moving a unit positive charge from infinity to that point against the direction of electric field.*

Therefore, if W is the work done in bringing a positive test charge q_0 from infinity to a point then potential at that point is

$$V \ = \ \frac{W}{q_0}$$

If W is in joule, q_0 is in Coulomb then potential V is in volt $\left(1 \text{ volt} = \dfrac{1 \text{ joule}}{1 \text{ Coulomb}}\right)$.

Consider a point charge +q placed at the point origin O. Let us calculate its electric potential at a point A situated at a distance r_A from it as shown in Fig. 1.9.

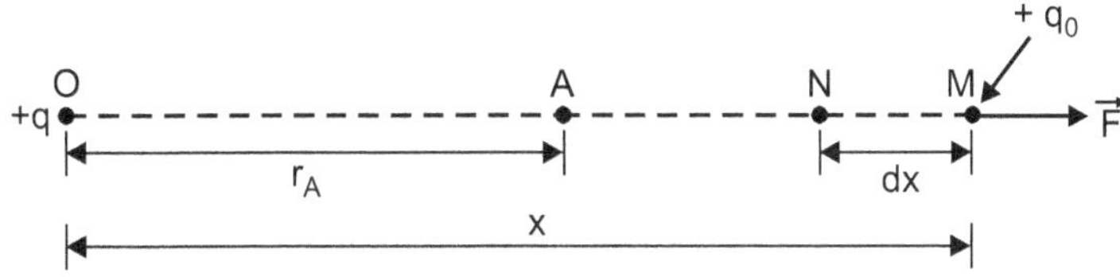

Fig. 1.9 : Electric potential due to a point charge

Imagine a test charge q_0 is placed at point M at a distant say x from origin O.

By Coulomb's law, the electrostatic force on charge q_0 is

$$F = \frac{1}{4\pi\varepsilon_0} \frac{qq_0}{x^2}$$

... (1.14)

The direction of force is shown in Fig. 1.9 and it acts away from the charge $+q$.

The work done (dw) in moving a test charge q_0 through small displacement $\overline{dx}$ against the direction of electric force is given by

$$dw = \overline{F} \cdot \overline{dx} = F\, dx \cos\theta$$

But, $\theta = 180°$ and $\cos 180° = -1$

$$\therefore \qquad dw = -F\, dx$$

... (1.15)

Using equation (1.15) in above equation (1.14), we get

$$dw = -\frac{1}{4\pi\varepsilon_0} \frac{qq_0}{x^2}\, dx$$

Now, the total work in moving a charge q_0 from point B (situated at distance r_B from point O) to point A will be

$$W = \int dw = -\int_{r_B}^{r_A} \frac{1}{4\pi\varepsilon_0} \frac{qq_0}{x^2}\, dx = -\frac{qq_0}{4\pi\varepsilon_0}\left[-\frac{1}{x}\right]_{r_B}^{r_A}$$

$$\therefore \qquad W = \frac{qq_0}{4\pi\varepsilon_0}\left[\frac{1}{r_A} - \frac{1}{r_B}\right]$$

Work done in moving a unit positive charge from infinity ($r_B = \infty$) to the point A (which is electric potential at point A) is

$$V = \frac{W}{q_0} = \frac{\dfrac{qq_0}{4\pi\varepsilon_0 r_A}}{q_0}$$

or

$$V = \frac{1}{4\pi\varepsilon_0} \frac{q}{r_A}$$

In general, if point A is considered at a distance r from charge $+q$,

then

$$V = \frac{1}{4\pi\varepsilon_0} \frac{q}{r}$$

This is the expression for potential of the charge $+q$ at a distance r from it.

1.3.2 Potential Energy of a Charge

- We have seen that electric field exists at all points around a charge q and every point in the space can be characterized with scalar quantity called the electric potential. The electric potential due to charge q at a point distant r from it is given by

$$V(\vec{r}) = \frac{1}{4\pi\varepsilon_0}\frac{q}{r}$$

- Electric potential at a point is the work done in bringing a unit positive charge infinity to that point. Instead of bringing a unit charge, if we bring a charge q' from infinity to that point, work done in bringing it is given by,

$$W = q' \times V(\vec{r}) = \frac{1}{4\pi\varepsilon_0}\frac{q'q}{r}$$

- This work done is termed as the potential energy of the charge q' in the field charge q.

    (i) If q and q' are of same nature, they are brought together against the force of repulsion and so potential energy of q' is positive.

    (ii) If q and q' are of opposite nature, the charges move under the effect of attractive forces and potential energy is then negative.

- From the above discussion, we can also say that *electric potential at any point in the electric field is the potential energy of unit positive charge placed at that point.*

1.3.3 Electrical Potential Energy of a System

- *The total work done in making an assembly of charges by bringing them from infinity their respective places is known as the electrical potential energy of a system.*

- To obtain an expression for potential energy of a system of charges, let us consider simple system of two charges q_1 and q_2 separated by a distance r_{12} as shown in Fig. 1.10.

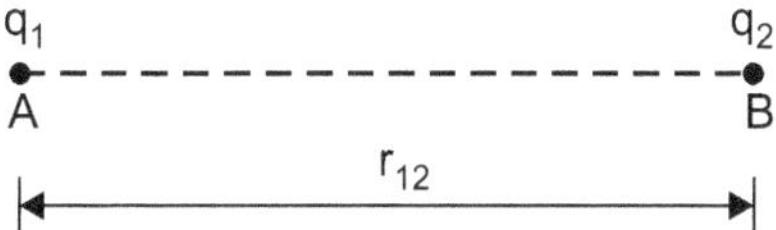

Fig. 1.10 : Potential energy of a system of two charges

- Let the charges q_1 and q_2 be brought from infinity to points A and B respectively one by one. When first charge q_1 is brought to point A, there was no electric field in the region and so no work has been done in bringing it to point A. After the charge q_1 had been placed at point A, an electric field is developed in the region. Now, the potential at point B, due to the charge q_1 is

$$V = \frac{1}{4\pi\varepsilon_0} \frac{q_1}{r_{12}}$$

- According to the definition of potential, this work is required to be done in order to bring a unit positive charge from infinity to point B. Therefore, the work done in bringing charge 'q_2' from infinity to point B will be

$$W_{12} = q_2 V = \frac{1}{4\pi\varepsilon_0} \frac{q_1 q_2}{r_{12}}$$

- This is the work done in bringing charge q_2 in the field of q_1. Thus, the electric potential energy U of the system of two charges q_1 and q_2 is

$$U = W_1 = \frac{1}{4\pi\varepsilon_0} \frac{q_1 q_2}{r_{12}}$$

- Now, suppose a third charge q_3 is brought from infinity to a point C, which is at a distance r_{13} from charge q_1 and r_{23} from charge q_2. The work required is

$$W = W_{13} + W_{23} = \frac{1}{4\pi\varepsilon_0} \frac{q_1 q_3}{r_{13}} + \frac{1}{4\pi\varepsilon_0} \frac{q_2 q_3}{r_{23}}$$

- Hence, the potential energy of the configuration of three charges is

$$U = W_{12} + W_{13} + W_{23}$$

$$U = \frac{1}{4\pi\varepsilon_0} \left[\frac{q_1 q_2}{r_{12}} + \frac{q_1 q_3}{r_{13}} + \frac{q_2 q_3}{r_{23}} \right]$$

- In general, if we have N different charges in any arrangement in space, then the total number of possible pairs for computing potential energy of the system are

(12, 13, 14, ... 1N), (23, 24, 25, ... 2N), (34, 35, ... 3N) and so on.

- Thus, a general expression for the potential energy of N different charges can be written as,

$$U = \frac{1}{2} \left(\frac{1}{4\pi\varepsilon_0} \right) \sum_{i=1}^{n} \sum_{j \neq i} \frac{q_i\, q_j}{r_{ij}} \qquad \ldots (1.16)$$

- The factor $\frac{1}{2}$ included in the above expression is due to the reason that the combinations $q_i\, q_j$ and $q_j\, q_i$ are to be counted as one.

- From above expression it is clear that the potential energy of a system of charges, assembled together depends only upon the magnitude of charges constituting the system and their positions with respect to each other.

1.4 Concept of Electric Field

- Between any two electric charges there always exists a force of attraction or repulsion. To explain how the charges exert a force on each other even though there is no physical contact between them, the concept of electric field (also called electrostatic field) has been introduced by Faraday. According to this concept -

   (i)   An electric charge produces an electric field in the surrounding space. When another charge is brought in this region of space, it experiences an electric force. The force is given by Coulomb's law.

   (ii) It is not a charge which exerts a force on another charge, but it is the field of a charge due to which another charge experiences an electric force. Thus an *electric field is said to exist in the region if another charge experiences an electric force in that region.*

- It must be noted that, the electric field has its own existence and is present even if there is no another charge to experience the force.

Electric Intensity ($\vec{E}$) :

The strength of electric field at any point in the region is described in terms of quantity called electric intensity or intensity of electric field. It is denoted by $\vec{E}$.

The intensity of an electric field at any point in an electric field is defined as the force acting on a unit positive charge placed at that point.

To determine electric intensity at a point in an electric field, imagine a small positive test charge q_0 is placed at that point. If $\vec{F}$ is the force acting on the test charge, the electric intensity at that point is given by

$$\vec{E} = \frac{\vec{F}}{q_0} \qquad \qquad \text{... (1.17)}$$

The electric intensity is a vector quantity and its direction is the direction of force experienced by positive test charge.

The magnitude of electric intensity is given by

$$|\vec{E}| = \frac{|\vec{F}|}{q_0}$$

Here a test charge q_o should be so small that it should not change the original electric field. Therefore, equation (1.17) in correct form is expressed as

$$\vec{E} = \lim_{q_o \to 0} \frac{\vec{F}}{q_o}$$

The S.I. unit of electric intensity is newton per coulomb (N/C). However, an equivalent unit for the electric intensity is volt/meter.

(a) Electric Intensity due to a Point Charge : (Nov. 10)

Consider a point charge '+q' situated in air at a point O. The presence of this charge gives rise to an electric field in its surrounding space. Let us find electric intensity at a point P, which is at a distance r from the charge q.

To find electric intensity ($\vec{E}$), imagine that a positive test charge q_o is placed at the point P. (Refer Fig. 1.11).

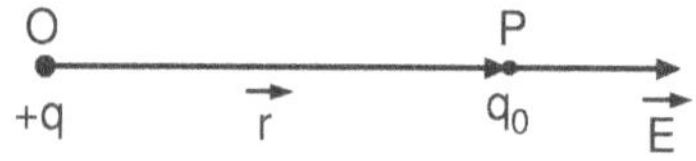

Fig. 1.11 : Electric intensity due to a point charge

The force on a charge q_o is $\vec{F} = \dfrac{1}{4\pi\varepsilon_o} \dfrac{qq_o}{r^2} \hat{r}$

where $\hat{r}$ is a unit vector along OP $\left(\hat{r} = \dfrac{\vec{r}}{|\vec{r}|} \right)$.

Electric intensity at point P is $\vec{E} = \dfrac{\vec{F}}{q_o}$

$$\therefore \quad \vec{E} = \frac{1}{4\pi\varepsilon_o} \frac{q}{r^2} \hat{r} \quad \text{or} \quad \vec{E} = \frac{1}{4\pi\varepsilon_o} \frac{q}{r^3} \vec{r} \quad \ldots (1.18)$$

This is the expression for electric intensity at point P. Since the charge q is positive, electric intensity is directed away from the charge. But if q is negative, the electric intensity is directed towards the charge. The magnitude of electric intensity at point P is

$$|\vec{E}| = \frac{1}{4\pi\varepsilon_o} \frac{q}{r^2} \quad \ldots (1.19)$$

(b) Electric Intensity due to Group of Point Charges :   (Oct. 15)

If we have a number of charges situated at different points in a given region, the electric intensity at any point in a given region due to this group of charges is calculated in following manner.

Place the test charge q_o, at the point where the electric intensity is to be calculated. The net force $\vec{F_o}$ acting on the charge q_o, from the n point charges $q_1, q_2, q_3, \ldots, q_n$ is

$$\vec{F_o} = \vec{F_{o1}} + \vec{F_{o2}} + \ldots + \vec{F_{on}}$$

where $\vec{F_{o1}}$ is the force acting on q_o due to q_1,

$\vec{F_{o2}}$ is the force acting on q_o due to q_2 and so on.

The resultant electric intensity at the given point (or at the position of the test charge) is

$$\vec{E} = \frac{\vec{F_o}}{q_o} = \frac{\vec{F_{o1}}}{q_o} + \frac{\vec{F_{o2}}}{q_o} + \ldots + \frac{\vec{F_{on}}}{q_o}$$

$$\vec{E} = \vec{E_1} + \vec{E_2} + \ldots + \vec{E_n}$$

This means that the vector sum of the electric intensities due to all the charges gives the resultant electric intensity at that point. Thus, we find that the electric field also obeys the superposition principle.

(c) Electric field due to continuous charge distribution :

So far we have studied forces and electric field due to point charges. It is possible to have a continuous distribution along a line, on a surface or in volume as shown in Fig. 1.12.

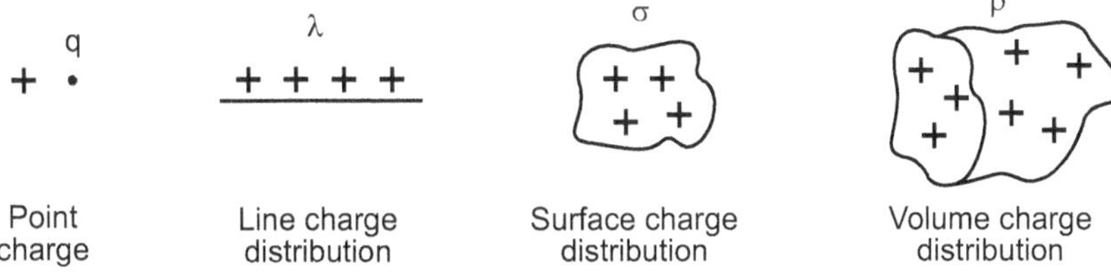

Fig. 1.12

When we deal with continuous charge distribution, charge on an object is expressed in terms of charge density, rather than a total charge.

(1) Volume charge density (ρ) : When the charge is distributed continuously in a certain volume, charge on object is expressed in terms of volume charge density (ρ). *Volume charge density is defined as charge per unit volume.*

If δq is the net charge enclosed by the volume δV situated at $\vec{r}$, then

$$\rho\,(\vec{r}\,)^* \;=\; \lim_{\delta V \to 0} \frac{\delta q}{\delta V}$$

S.I. unit of ρ is coulomb/(meter)^3 i.e. (C/m^3).

(2) Surface charge density (σ) : When the charge is distributed continuously over a surface, the charge on an object is expressed in terms of surface charge density (σ). *It is defined as charge per unit area.* If δq is the net charge on small surface element δS of negligible thickness situated at $\vec{r}$, then $\sigma(\vec{r}\,) = \lim_{\delta S \to 0} \dfrac{\delta q}{\delta S}$

S.I. unit of σ is coulomb/(meter)^2 i.e. (C/m^2).

(3) Line charge density (λ) : If charges are continuously distributed along a length or line, then charge on an object is expressed in terms of line charge density (λ).

It is defined as charge per unit length.

$$\lambda\,(\vec{r}\,) \;=\; \lim_{\delta l \to 0} \frac{\delta q}{\delta l}$$

where, δl is small length element situated at $\vec{r}$ and carries a charge dq upon it.

S.I. unit of λ is coulomb/meter.

The total charge in a finite region of space having charge distributed on surface and in volume is obtained by integration of charge densities over the region i.e.

$$q \;=\; \int_V \rho\,dV + \int_S \sigma\,dS$$

In order to calculate the electric intensity or potential at any point, due to continuous charge distribution, the charge is imagined to be divided into infinitesimal elements dq. Considering any elementary charge dq as a point charge, electric intensity at a given point is determined. Appropriate integration gives the total intensity at that point.

---

* *If charges are uniformly distributed, ρ is not a function of $\vec{r}$ and is expressed as,*
$$\rho \;=\; \lim_{\delta V \to 0} \frac{\delta q}{\delta V}$$

1.5 Concept of Electric Flux

- The word "flux" comes from a latin word meaning "to flow". We can consider the flux of a vector field to be a measure of flow or penetration of the field vectors through an imaginary small surface element dS situated in the field.

- In electrostatics, *the number of electrical lines of force passing normally through a given surface area drawn in an electric field is called flux of electric field or electric flux through given area.* It can be determined in the following manner.

- Consider a small surface area dS around a point P situated in an electric field as shown in Fig. 1.13. Let $\vec{E}$ be the electric intensity at point P. Since the area dS is infinitely small, the electric intensity at every point of this area can be supposed to be a constant and is equal to $\vec{E}$.

- The flux of electric field ($d\phi_E$) through the area is defined as the *product of the area and the component of electric intensity perpendicular to the area.*

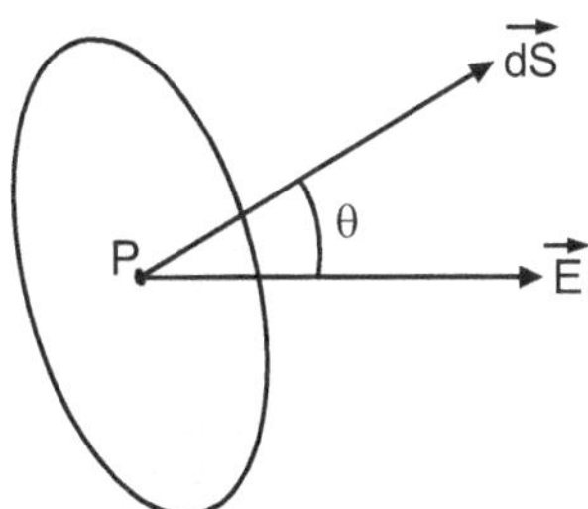

Fig. 1.13 : Flux of electric field through an infinitesimally small area dS

From Fig. 1.13, the component of electric intensity perpendicular to area dS is E cos θ.

$$\therefore \quad d\phi_E = E \cos\theta \, dS$$

The area dS can be represented vectorially $\vec{dS}$ drawn perpendicular to the area. Thus, above equation may be written as

$$d\phi_E = \vec{E} \cdot \vec{dS}$$

The flux of electric field is a scalar quantity and its SI unit is $N \cdot m^2/C$.

1.6 Gauss's Law in Electrostatics

- Gauss's theorem gives the relation between the electric flux through any closed hypothetical surface (called as Gaussian surface) and the total charge enclosed by the surface.

Statement : *It states that the total electric flux ϕ_E through any closed surface is equal to $\dfrac{q_{enc}}{\varepsilon_o}$, where q_{enc} is the total charge inside the closed surface. i.e.*

$$\phi_E = \frac{q_{enc}}{\varepsilon_o}$$

Proof : To prove Gauss's law, consider any closed surface S as shown in Fig. 1.14.

Let charges q_1, q_2, q_3 ...... be situated at points A, B, C ... respectively inside the closed surface 'S'.Imagine a small element of surface area dS around any point P as shown in Fig. 1.14.

Let us first consider a charge q_1. Let the distance of point P from point A be r.

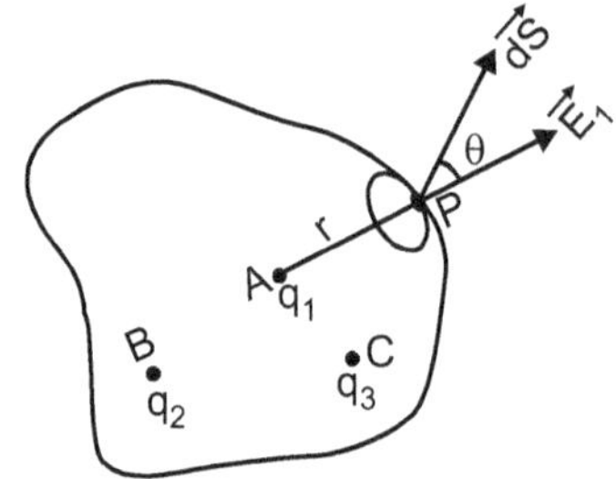

Fig. 1.14 : Gauss's law

The electric field intensity at a point P due to charge q_1 is given by

$$E_1 = \frac{1}{4\pi\varepsilon_0}\frac{q_1}{r^2}$$

The electric field intensity E_1 is directed along AP as shown in Fig. 1.14. The small electric flux $d\phi_1$ passing through the small surface area dS due to charge q_1 is given by $d\phi_1 = \bar{E}_1 \cdot d\bar{S} = E_1\, dS \cos\theta$

where θ is the angle between direction of $\bar{E}_1$ and normal to the surface element dS.

The total electric flux ϕ_1 passing through the whole surface S due to charge q_1 is given by $\phi_1 = \oint d\phi_1$

$$\phi_1 = \oint E_1\, dS \cos\theta$$

But,

$$E_1 = \frac{1}{4\pi\varepsilon_0}\frac{q_1}{r^2}$$

$$\therefore \quad \phi_1 = \frac{1}{4\pi\varepsilon_0} q_1 \oint \frac{dS \cos\theta}{r^2}$$

Let, $\dfrac{dS \cos\theta}{r^2} = dw$ represent small solid angle subtended by element of area dS at the location of charge q_1.

$$\therefore \qquad \phi_1 = \frac{q_1}{4\pi\varepsilon_0} \oint dw \qquad\qquad \dots (1.20)$$

In above equation, $\oint dw$ is the total solid angle surrounding the point A. But from solid geometry, the total solid angle surrounding any point in space is 4π steradians.

$$\therefore \qquad \oint dw = 4\pi$$

Using above equation in equation (1.20), we get

$$\phi_1 = \frac{q_1}{4\pi\varepsilon_0} 4\pi \ \text{ or } \ \phi_1 = \frac{q_1}{\varepsilon_0}$$

Similarly, electric flux ϕ_2, ϕ_3 ... due to charges q_2, q_3 ... respectively in the space enclosed by S will be $\phi_2 = \dfrac{q_2}{\varepsilon_0}$, $\phi_3 = \dfrac{q_3}{\varepsilon_0}$

Hence, the total electric flux passing through the closed surface due to all charges enclosed by it, can be expressed as

$$\phi = \phi_1 + \phi_2 + \phi_3 + \dots = \frac{q_1}{\varepsilon_0} + \frac{q_2}{\varepsilon_0} + \frac{q_3}{\varepsilon_0} + \dots$$

$$= \frac{1}{\varepsilon_0}(q_1 + q_2 + q_3 + \dots)$$

$$\phi = \frac{q_{enc}}{\varepsilon_0}$$

Thus, if q_{enc} is the total charge enclosed by the closed surface then

$$\oint \vec{E} \cdot d\vec{S} = \frac{q_{enc}}{\varepsilon_0}$$

This equation is known as integral form of Gauss's law.

Gauss's law is the basic theorem of electrostatics. This law is always true, but is not always useful. Gauss's law provides an easy method of finding $\vec{E}$ due to a point charge or due to a given charge distribution. The usefulness of Gauss's law depends on our ability to evaluate $\oint \vec{E} \cdot d\vec{S}$.

We can easily evaluate $\oint \vec{E} \cdot d\vec{S}$ if the given charge distribution is uniform and symmetric. It means that Gauss's law is useful, if the charge distribution is uniform and symmetric. It must be noted that, if the charge distribution is not symmetric and uniform, then Gauss's law is still obeyed, but we cannot use it to determine $\vec{E}$. In such a case, Coulomb's law may be used to determine $\vec{E}$.

Gauss's law can be used to find the electric field intensity in a situation where there is uniform and symmetrical distribution of charge. We shall now consider some applications of Gauss's law.

Illustrative Examples

Example 1 : Electric Intensity at a Point due to Uniformly Charged Non-conducting Sphere.

Case (i) : At an external point : Consider a sphere of non-conducting material of radius 'a' (Refer Fig. 1.15). The charge +q is distributed uniformly throughout a volume (V) of the sphere. Let P be any point outside the sphere at a distance 'R' from its centre O.

To find the electric intensity at a point P, imagine a Gaussian surface S in the form of concentric sphere of radius R. The electric intensity is radially outward and it has the same magnitude at all points on the Gaussian surface. The angle between $\vec{E}$ and normal to any surface element of area dS on the Gaussian surface is zero.

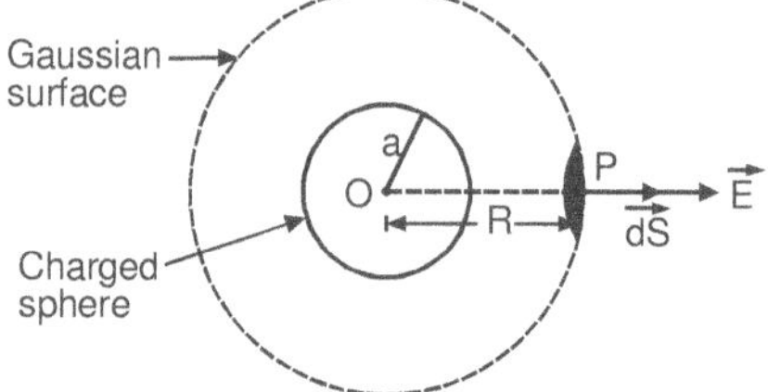

Fig. 1.15 : A uniformly charged sphere

Flux passing through small surface element $\vec{dS}$ is

$$d\phi = \vec{E} \cdot \vec{dS} = E \, dS \cos\theta$$

But
$$\theta = 0$$
$$\therefore \quad d\phi = E \, dS$$

Total flux through Gaussian surface is

$$\phi = \int d\phi = \oint E \, dS$$

$$\phi = E \oint dS \quad \text{or} \quad \phi = E \, 4\pi R^2$$

According to Gauss's theorem, $\phi = \dfrac{q_{enc}}{\varepsilon_o}$

where q_{enc} is total charge enclosed by Gaussian surface.

$$\therefore \qquad E\,4\pi R^2 \;=\; \frac{q}{\varepsilon_o} \qquad\qquad (\because q_{enc} = q)$$

$$\text{or} \qquad E \;=\; \frac{1}{4\pi\varepsilon_o}\frac{q}{R^2}$$

This is the expression for electric intensity, due to a charged sphere at an external point P, distant R from the centre of sphere. Since charge on the sphere is positive, $\vec{E}$ is directed away from the charge and is along the line joining given point and centre O (shown in Fig. 1.15).

This equation shows that for the points outside the charged sphere, it behaves as if all the charges are concentrated at the centre of the sphere.

Case (ii) : At a point on the surface of sphere :

If the point P is on the surface of sphere, then R = a and we get,

$$E \;=\; \frac{1}{4\pi\varepsilon_o}\frac{q}{a^2}$$

Case (iii) : At an internal point :

Let us find electric intensity at a point P' which is situated inside the charged sphere at a distance r from the centre O (Refer Fig. 1.16). To find the electric intensity at point P', imagine a Gaussian spherical surface of radius r.

The electric intensity is radially outward and has the same magnitude at all points on the Gaussian surface. The angle between $\vec{E}$ and $\vec{dS}$ at any point on Gaussian surface is zero.

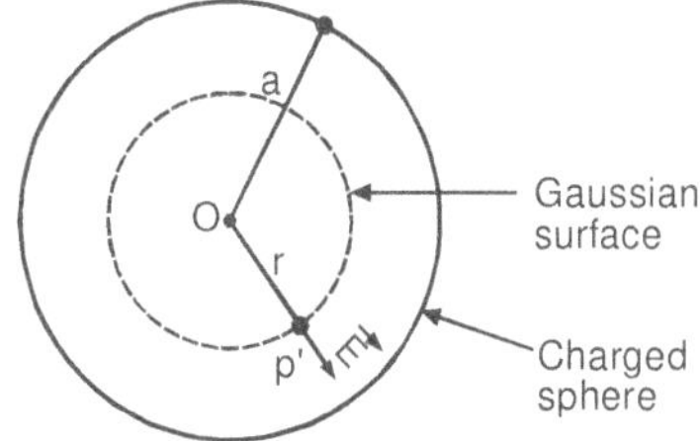

Fig. 1.16 : Uniformly charged sphere

Flux passing through small surface element dS is

$$d\phi \;=\; \vec{E} \cdot \vec{dS} = EdS\cos\theta$$

But $\qquad\qquad \theta = 0$

$\therefore \qquad\qquad d\phi = E\,dS$

Total flux passing through Gaussian surface is

$$\phi \;=\; \int d\phi = \oint E\,dS$$

$$\phi \;=\; E\oint dS \quad\text{or}\quad \phi = E\,4\pi r^2 \qquad\qquad (\because S = 4\pi r^2)$$

According to Gauss's theorem, $\phi = \dfrac{q_{enc}}{\varepsilon_o}$ ... (1.21)

where q_{enc} is charge enclosed by Gaussian surface.

Let us find charge enclosed by Gaussian surface.

$$\text{Volume charge density } (\rho) = \frac{\text{Charge (dq)}}{\text{Volume (dV)}}$$

where dq is the charge carried by volume dV.

Charge enclosed by the Gaussian surface = ρ.

Volume enclosed by Gaussian surface, $q_{enc} = \rho \cdot \dfrac{4}{3}\pi r^3$ ... (1.22)

Substituting equation (1.22) in equation (1.21), we get

$$\phi = \frac{\rho \dfrac{4}{3}\pi r^3}{\varepsilon_o}$$

But $$\phi = E\, 4\pi r^2$$

$\therefore$ $$E\, 4\pi r^2 = \frac{\rho \dfrac{4}{3}\pi r^3}{\varepsilon_o}$$

$$E = \frac{\rho}{3\varepsilon_o} r$$ ... (1.23)

This is the expression for electric intensity, due to a charged sphere at an internal point, distance r from the centre of sphere. E is directed away from the charge and is along the line joining given point and centre O.

This equation shows that electric intensity at a point inside the sphere is directly proportional to distance of that point from the centre O.

Fig. 1.17 shows variation of E as a function of radial distance (x).

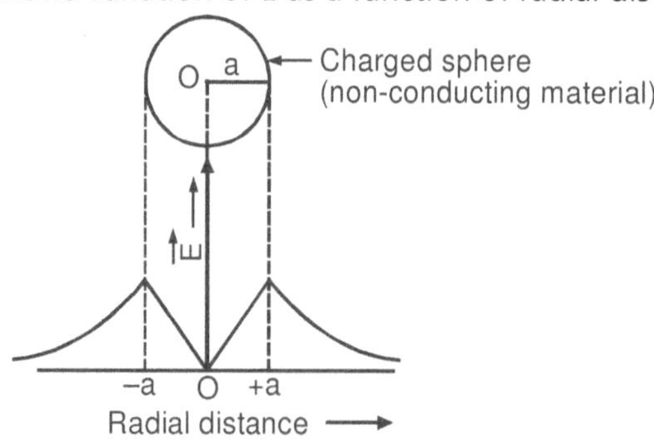

Fig. 1.17

Example 2 : Electric Intensity due to Charged Plane Conductor.

Consider a conductor which is uniformly charged over its surface as shown in Fig. 1.18. The electric intensity at any point outside the conductor is always perpendicular to the surface. If this were not so, the electric intensity will have a component parallel to the surface and it will produce movement of charges along the surface. But as the charges are at rest, the electric intensity outside the surface must be everywhere perpendicular to the surface.

Let us find the electric intensity at a point P which is close to the surface using Gauss's law.

Imagine a Gaussian surface in the form of cylinder having small cross-sectional area dS, drawn with its axis perpendicular to the conductor having one end B inside the conductor and the other end A outside the conductor.

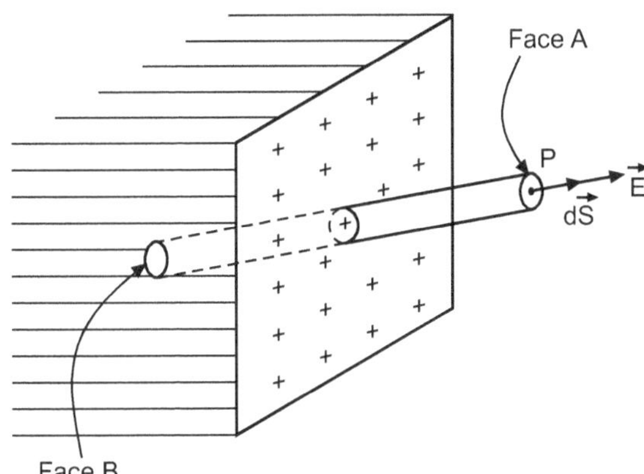

Fig. 1.18 : Electric intensity close to the charged plane conductor

As there is no field inside the conductor (i.e. E = 0), the flux across the face B will be zero.

As $\vec{E}$ will be parallel to the axis of the cylinder (outside the region), there will be no electric flux across the curved surface of the cylinder.

Hence, the total flux over the cylinder is the flux passing through the end surface A. If the surface area of face A is taken as dS, then

Flux through the surface area dS $= \bar{E} \cdot \bar{dS}$

$$\phi = E\, dS \cos \theta$$

$$\phi = E\, dS \qquad\qquad (\because \theta = 0)$$

According to Gauss's theorem, $\phi = \dfrac{q_{enc}}{\varepsilon_0}$

If σ is the surface charge density, then charge on area dS is $q_{enc} = \sigma \, dS$

$$\therefore \qquad E \, dS = \frac{\sigma \, dS}{\varepsilon_0}$$

$$E = \frac{\sigma}{\varepsilon_0}$$

This is the expression for magnitude of the electric intensity at any point just outside the metallic conductor.

For a charged conductor placed in a medium of dielectric constant K,

we have $\qquad\qquad E = \dfrac{\sigma}{K \, \varepsilon_0}$

Example 3 : Electric field intensity due to infinitely long charged wire.

Electric Field due to a Line Charge :   **(Oct. 15)**

Consider an uniformly charged infinitely long thin wire. Let AB be a part of a thin wire having a charge q per unit length (Refer Fig. 1.19). To find the electric intensity at a point P situated at a distance r from the wire, imagine a cylinder of radius r and length l having the given wire as its axis. The imaginary cylinder is a Gaussian surface and point P is on its surface as shown in Fig. 1.19.

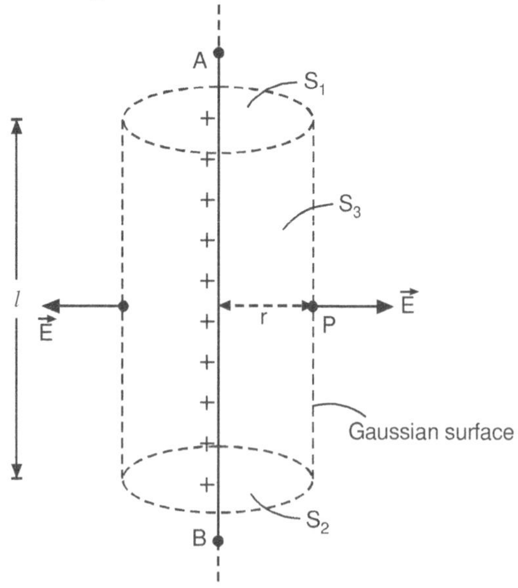

Fig. 1.19 : Uniformly charged thin wire AB
surrounded by cylindrical Gaussian surface

The cylindrical Gaussian surface has two types of surfaces.

(a) Flat surfaces S_1 and S_2 : As there is no electric field perpendicular to the flat faces S_1 and S_2, the electric flux through these faces is zero.

(b) Curved surface S_3 : The electric intensity at every point of the curved surface of the Gaussian surface is same and it is perpendicular to the surface. Therefore, outward flux through the Gaussian surface is

$$\phi_E = \oint_S \vec{E} \cdot \vec{dS} = \oint_S E \, dS \cos\theta$$

But, $\theta = 0$

$\therefore$ $\phi_E = E \int dS = E \times (\text{Area of curved surface})$

$$\phi_E = E \, 2\pi r l \qquad \qquad \text{... (1.24)}$$

According to Gauss's theorem,

$$\phi_E = \frac{q_{enc}}{\varepsilon_o}$$

where, q_{enc} is the total charge enclosed by the closed surface.

The total charge enclosed by the Gaussian surface is ql.

$\therefore$ $$\phi_E = \frac{ql}{\varepsilon_o} \qquad \qquad \text{... (1.25)}$$

From equations (1.24) and (1.25), we get

$$E \times 2\pi r l = \frac{ql}{\varepsilon_o} \quad \text{or} \quad E = \frac{q}{2\pi\varepsilon_o r}$$

This is the expression for magnitude of electric intensity at a distance r from the uniformly charged wire having a charge q per unit length. The direction of electric intensity is perpendicular to the axis of wire.

Solved Problems

Problem 1.4 : *Calculate the force between two balls each having a charge of 12 μC and are 8 cm apart.*   **(April 16, Oct. 16)**

Solution : Given : (i) Charge on the first ball, $q_1 = 12 \ \mu C = 12 \times 10^{-6}$ C

(ii) Charge on the second ball, $q_2 = 12 \ \mu C = 12 \times 10^{-6}$ C

(iii) Distance between two charges, $r = 8 \ cm = 8 \times 10^{-2}$ m

Formula : The magnitude of force between two balls is

$$F = \frac{1}{4\pi\varepsilon_o} \frac{q_1 q_2}{r^2} = 9 \times 10^9 \times \frac{12 \times 10^{-6} \times 12 \times 10^{-6}}{(8 \times 10^{-2})^2}$$

$$= \frac{9 \times 144}{64} \times 10^1 = \textbf{202.5 N} \qquad \qquad \textbf{... Ans.}$$

The force between two balls is 202.5 N. The direction of the force is as shown in Fig. 1.20.

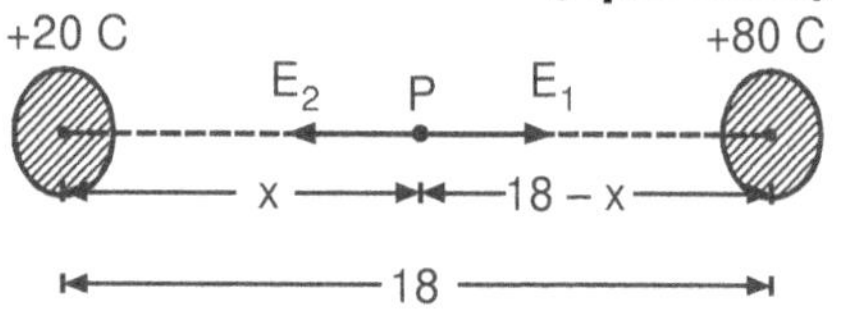

Fig. 1.20

$$\left|\vec{F_{12}}\right| = \left|\vec{F_{21}}\right| = \textbf{202.5 N} \qquad \textbf{... Ans.}$$

Problem 1.5 : *Two spheres of charges +20 and 80 coulomb are placed 18 cm apart. Find the position of the point between them where the intensity is zero.*

(April 2015)

Solution : Fig. 1.21 shows two charges +20 C and 80 C separated by a distance 18 cm.

Let P be a point at a distance x from +20 C in between the two charges, where the electric intensity is zero.

Fig. 1.21

The magnitude of $\vec{E}$ is given by, $E_1 = \dfrac{1}{4\pi\varepsilon_o}\dfrac{q}{r^2}$

Therefore, the electric intensity at point P due to charge +20 C is

$$E_1 = \dfrac{1}{4\pi\varepsilon_o}\dfrac{20}{x^2}$$

E_1 is directed from left to right as shown in Fig. 1.23.

The electric intensity at point P due to charge 80 C is

$$E_2 = \dfrac{1}{4\pi\varepsilon_o}\dfrac{80}{(18-x)^2}$$

E_2 is directed from right to left as shown in Fig. 1.21.

For electric intensity to be zero at point P,

$$E_1 = E_2$$

$$\therefore \quad \dfrac{1}{4\pi\varepsilon_o}\dfrac{20}{x^2} = \dfrac{1}{4\pi\varepsilon_o}\dfrac{80}{(18-x)^2}$$

$$\text{or} \quad \dfrac{1}{x^2} = \dfrac{4}{(18-x)^2}$$

Taking square root of both sides, we get

$$\dfrac{1}{x} = \dfrac{2}{18-x}$$

$$\text{or} \quad 18 - x = 2x$$

$$\therefore \quad x = \textbf{6 cm} \qquad \textbf{... Ans.}$$

Thus, distance of point P, where the electric intensity is zero, is at a distance 6 cm from the charge +20 C.

Problem 1.6 : *Find the magnitude and direction of an electric intensity at a point distant z above the mid-point between two equal charges '+q', at a distance d apart (Refer Fig. 1.22).*

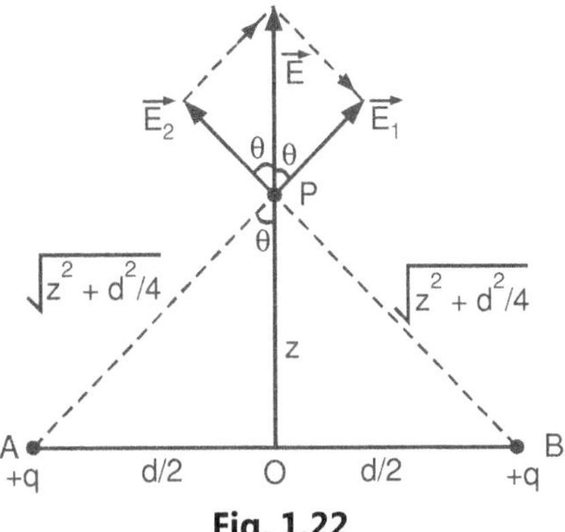

Fig. 1.22

Solution : The magnitude of electric intensity $|\vec{E_1}|$ at point P due to charge +q situated at point A is

$$|\vec{E_1}| = \frac{1}{4\pi\varepsilon_o} \frac{q}{\left(\sqrt{z^2 + \dfrac{d^2}{4}}\right)^2} \quad \text{or} \quad E_1 = \frac{1}{4\pi\varepsilon_o} \frac{q}{\left(z^2 + \dfrac{d^2}{4}\right)}$$

The component of E_1 along z-axis is $E_{1z} = E_1 \cos\theta$.

$$\therefore \qquad E_{1z} = \frac{1}{4\pi\varepsilon_o} \frac{q}{\left(z^2 + \dfrac{d^2}{4}\right)} \cos\theta$$

The magnitude of electric intensity $|\vec{E_2}|$ at point P due to charge +q situated at point B is

$$E_2 = \frac{1}{4\pi\varepsilon_o} \frac{q}{\left(z^2 + \dfrac{d^2}{4}\right)}$$

The component of E_2 along z-axis is

$$E_{2z} = \frac{1}{4\pi\varepsilon_o} \frac{q}{\left(z^2 + \dfrac{d^2}{4}\right)} \cos\theta$$

The resultant intensity at point P due to charges +q situated at A and +q at B is

$$E = E_{1z} + E_{2z} = \frac{2q\cos\theta}{4\pi\varepsilon_o\left(z^2 + \dfrac{d^2}{4}\right)}$$

Note that the component perpendicular to z-axis cancels out.

But, from Fig. 1.22,

$$\cos\theta = \frac{z}{\sqrt{z^2 + \dfrac{d^2}{4}}}$$

$$\therefore \quad E = \frac{2qz}{4\pi\varepsilon_o\left(z^2 + \dfrac{d^2}{4}\right)^{3/2}}$$

The resultant intensity at P is directed along the positive direction of z-axis. If z >> d, then electric intensity at P is given by,

$$E \approx \frac{2qz}{4\pi\varepsilon_o\,z^3} \quad \text{or} \quad E \approx \frac{1}{4\pi\varepsilon_o}\frac{2q}{z^2} \qquad \textbf{... Ans.}$$

Thus, when z >> d, the charges appear to be concentrated at centre O.

Problem 1.7 : *Three point charges of +1 μC, +2 μC and +3 μC are at the vertices of an equilateral triangle of side 10 cm. Find the magnitude of the resultant force acting on the + 3 μC charge.*

Solution : Given : $q_1 = 1\ \mu C = 1 \times 10^{-6}$ C, $q_2 = 2\ \mu C = 2 \times 10^{-6}$ C,

$q_3 = 3\ \mu C = 3 \times 10^{-6}$ C, r = 10 cm = 10×10^{-2} meter

These three charges q_1, q_2, q_3 are at the vertices of an equilateral triangle ABC as shown in Fig. 1.23.

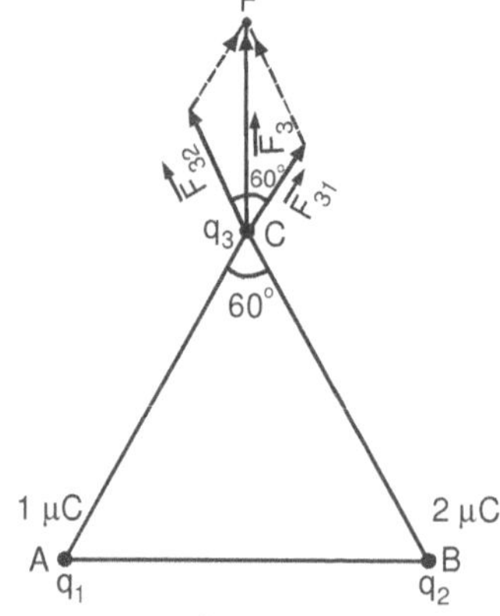

Fig. 1.23

The magnitude of force between two charges q_1 and q_3 separated by a distance r is given by $F = \dfrac{1}{4\pi\varepsilon_o}\dfrac{q_1\,q_3}{r^2}$ where $\dfrac{1}{4\pi\varepsilon_o} = 9 \times 10^9$ N-m²/C²

(a) The magnitude of force on q_3 due to q_1 is

$$F_{31} = \frac{1}{4\pi\varepsilon_o}\frac{q_1\,q_3}{|\vec{r}_{13}|^2} = 9 \times 10^9 \times \frac{1 \times 10^{-6} \times 3 \times 10^{-6}}{(10 \times 10^{-2})^2} = 2.7 \text{ newton}$$

(b) The magnitude of force on q_3 due to q_2 is

$$F_{32} = \frac{1}{4\pi\varepsilon_o} \frac{q_2 \, q_3}{|\vec{r}_{23}|^2}$$

$$|\vec{r}_{23}| = r = 10 \times 10^{-2} \text{ m}$$

$$\therefore \quad F_{32} = 9 \times 10^9 \times \frac{2 \times 10^{-6} \times 3 \times 10^{-6}}{(10 \times 10^{-2})^2} = 5.4 \text{ newton}$$

The directions of $\vec{F}_{32}$ and $\vec{F}_{31}$ are shown in Fig. 1.23.

The magnitude of resultant force on charge $q_3 = 3 \ \mu C$ can be calculated by using the law of parallelogram of vectors.

$$(R = \sqrt{P^2 + Q^2 + 2 \ PQ \cos \theta})$$

$$F_3 = \sqrt{(2.7)^2 + (5.4)^2 + 2 \times (2.7) \times (5.4) \cos 60}$$

$$\left(\because P = |\vec{F}_{31}| \text{ and } Q = |\vec{F}_{32}| \right)$$

$$F_3 = \textbf{7.14 N} \qquad\qquad \textbf{... Ans.}$$

The resultant force F_3 is directed along CF.

Problem 1.8 : *An electric flux of 6×10^3 Nm^2 C^{-1} is found to be linked with a sphere due to some charge inside it. Calculate the magnitude of charge inside the sphere. (Given : $\varepsilon_o = 8.85 \times 10^{-12}$ C^2/Nm^2).*

Solution : Given : (i) $\phi_E = 6 \times 10^3$ Nm^2 C^{-1}

(ii) $\qquad\qquad \varepsilon_0 = 8.85 \times 10^{-12}$ C^2/Nm^2

Formula : According to Gauss's theorem,

$$\phi_E = \frac{q_{enc}}{\varepsilon_0}$$

$$\therefore \quad q_{enc} = \varepsilon_o \, \phi_E = 8.85 \times 10^{-12} \times 6 \times 10^3 \text{ C}$$

$$= \textbf{5.3} \times \textbf{10}^{\textbf{-8}} \textbf{C} \qquad\qquad \textbf{... Ans.}$$

Therefore, total charge inside the sphere is 5.3×10^{-8} C.

Problem 1.9 : *A charge of 12 nano-coulombs is situated inside a cube. Calculate the electric flux through one of the faces of the cube.*

(Given : $\varepsilon_o = 8.85 \times 10^{-12}$ C^2/Nm^2).

Solution : Given :

(i) Total charge enclosed by the cube, $q_{enc} = 12 \times 10^{-9}$ C

Formula : According to Gauss's theorem,

$$\phi_E = \frac{q_{enc}}{\varepsilon_0} = \frac{12 \times 10^{-9}}{8.85 \times 10^{-12}} = 1.36 \times 10^3 \text{ Nm}^2\text{/C}$$

This electric flux is distributed equally over all the faces of cube which are of equal areas. Therefore, flux linked with each face area

$$= \frac{1.36 \times 10^3}{6} = 225.9 \ \text{Nm}^2/\text{C} \qquad \ldots \textbf{Ans.}$$

Problem 1.10 : *Charges of 12 nanocoulomb, −12 nanocoulomb and 24 nanocoulomb are placed at the three corners P, R and S of a square of side 4 cm as shown in Fig. 1.24. Find the electric intensity at the fourth corner Q.*

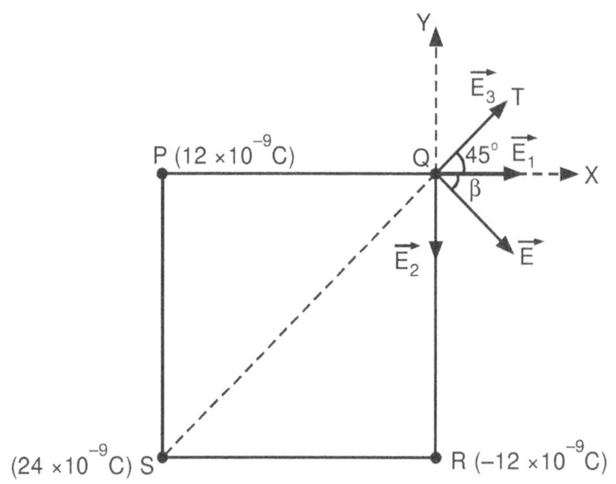

Fig. 1.24

Solution : Let E_1, E_2 and E_3 be electric intensities produced at point Q by charges situated at P, R and S respectively. $|\vec{E}| = \dfrac{1}{4\pi\varepsilon_o} \dfrac{q}{d^2}$

From Fig. 1.24, we get $E_1 = \dfrac{1}{4\pi\varepsilon_o} \dfrac{12 \times 10^{-9}}{(0.04)^2}$

$$\text{or} \qquad E_1 = \frac{9 \times 10^9 \times 12 \times 10^{-9}}{(0.04)^2}$$

$$= 6.75 \times 10^4 \ \text{N/C directed along QX.}$$

$$E_2 = \frac{1}{4\pi\varepsilon_o} \frac{12 \times 10^{-9}}{(0.04)^2}$$

$$\text{or} \qquad E_2 = \frac{9 \times 10^9 \times 12 \times 10^{-9}}{(0.04)^2}$$

$$= 6.75 \times 10^4 \ \text{N/C directed along QR}$$

$$\text{and} \qquad E_3 = \frac{1}{4\pi\varepsilon_o} \frac{24 \times 10^{-9}}{32 \times 10^{-4}}$$

$$\text{or} \qquad E_3 = \frac{9 \times 10^9 \times 24 \times 10^{-9}}{32 \times 10^{-4}}$$

$$= 6.75 \times 10^4 \text{ N/C directed along QT}$$

The component of E_3 along QX is

$$E_3 \cos 45° = 4.77 \times 10^4 \text{ N/C}$$

Total electric intensity along QX is

$$E_x = E_1 + E_3 \cos 45°$$

$$= 6.75 \times 10^4 + 4.77 \times 10^4 = 11.52 \times 10^4 \text{ N/C}$$

The component of E_3 along QY is

$$E_3 \sin 45° = 4.77 \times 10^4 \text{ N/C}$$

Total electric intensity along QR is

$$E_y = E_2 - E_3 \sin 45 = 6.75 \times 10^4 - 4.77 \times 10^4$$

$$E_y = 1.98 \times 10^4 \text{ N/C}$$

The resultant intensity at point Q is

$$E = \sqrt{E_x^2 + E_y^2}$$

$$E = \sqrt{(11.52 \times 10^4)^2 + (1.98 \times 10^4)^2}$$

$$= \sqrt{136.63} \times 10^4$$

$$E = \mathbf{11.68 \times 10^4 \text{ N/C}} \qquad \textbf{... Ans.}$$

This is the resultant electric intensity at point Q, due to three charges as shown in Fig. 1.24.

The angle made by resultant electric intensity with X-axis is

$$\tan \beta = \frac{E_y}{E_x} = \frac{1.98 \times 10^4}{11.52 \times 10^4}$$

$$\tan \beta = 0.171$$

$$\therefore \qquad \beta = \mathbf{9.70°} \qquad \textbf{... Ans.}$$

Problem 1.11 : *Using Gauss's theorem, obtain an expression for electric intensity at any point due to uniformly charged conducting sphere.*

Solution : Case (i) : At an external point : Consider a sphere of conducting material of radius 'a' (Fig. 1.25) and centre O. Since the sphere is of conducting material, charge resides on its surface only and no charge will be present at the interior of the sphere. Let charge +q be a total charge uniformly distributed on the surface of the sphere.

Let P be any point outside the sphere at a distance R from its centre 'O'.

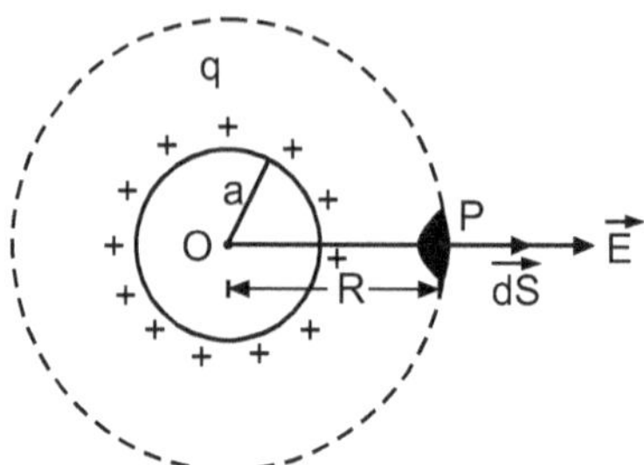

Fig. 1.25 : Uniformly charged conducting sphere

To find the electric intensity at a point P, imagine a Gaussian surface S in the form of concentric sphere of radius R. The electric intensity is radially outward and it has the same magnitude at all points on the Gaussian surface. The angle between $\vec{E}$ and normal to any surface element of area dS on the Gaussian surface is zero.

Flux passing through small surface element $\vec{dS}$ is

$$d\phi = \vec{E} \cdot \vec{dS} = E\, dS \cos\theta$$

But

$$\theta = 0$$

$\therefore$

$$d\phi = E\, dS$$

Total flux through Gaussian surface is $\phi = \int d\phi = \oint E\, dS$

$$\phi = E \oint dS$$
$$\phi = E\, 4\pi R^2$$

According to Gauss's theorem, $\phi = \dfrac{q_{enc}}{\varepsilon_0}$

But,

$$q_{enc} = q$$

$\therefore$

$$E\, 4\pi R^2 = \frac{q}{\varepsilon_0} \quad \text{or} \quad E = \frac{1}{4\pi\varepsilon_0} \frac{q}{r^2}$$

This is the expression for the electric intensity due to a charged sphere at an external point P distant R from the centre of the sphere.

Case (ii) : At a point on the surface of the sphere : If the point P is on the surface of a sphere, then R = a and we get

$$E = \frac{1}{4\pi\varepsilon_0} \frac{q}{a^2}$$

Case (iii) : At an internal point : Let us find E at point P' which is situated inside the charged sphere at a distance r from the centre O (Fig. 1.26).

We have seen that the charge given to a conductor resides on its surface only. No charge can be present in the interior of the conductor. Therefore, if we consider Gaussian surface S' of radius r lying completely inside the charged conducting sphere (Fig. 1.26), then the charge enclosed by the Gaussian surface will be zero. Hence from Gauss's theorem, the total electric flux passing the surface S' will be

$$\phi = \frac{q_{enc}}{\varepsilon_0} = 0$$

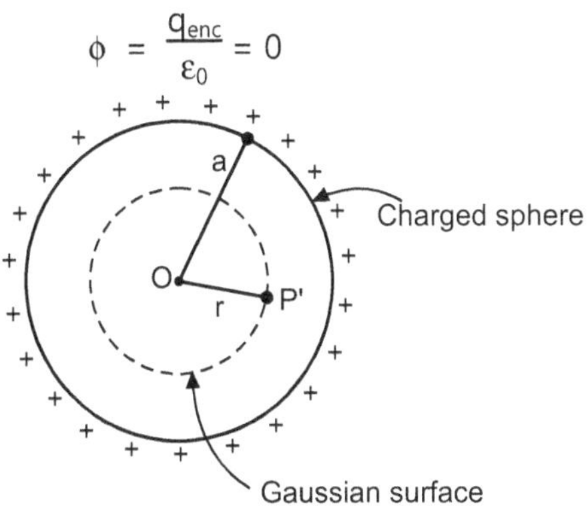

Fig. 1.26

The total flux through Gaussian surface is given by

$$\phi = \oint E \cdot dS' = E\, 4\pi r^2$$

But, $\phi = 0$

$\therefore$ $E\, 4\pi r^2 = 0 \text{ or } E = 0$

Thus, at any point in the interior of the spherical conductor, the electric field intensity is zero.

Problem 1.12 : *The electric field intensity at a point at a distance of 1 m from the centre of a charged sphere of radius 30 cm is 10^4 N/C. Find the surface charge density on the surface of sphere. The sphere is situated in air.*   **(April 16, Oct. 15)**

Solution : Given : r = 1 m, Radius of sphere, a = 30 cm = 0.30 m.

Electric intensity, E = 10^4 N/C.

To calculate : Surface charge density σ.

Formula : The magnitude of electric field intensity at a distance r from the centre of sphere is given by, $E = \dfrac{1}{4\pi\varepsilon_0}\, \dfrac{q}{r^2}$

$\therefore$ $q = E \times 4\pi\varepsilon_0\, r^2 = 10^4 \times 4\pi \times 8.85 \times 10^{-12} \times (1)^2$

                    $= 8.85 \times 4\pi \times 10^{-8}$ C

Surface charge density is given by

$$\sigma = \frac{\text{Charge on sphere}}{\text{Area of sphere}}$$

$$= \frac{q}{4\pi a^2} = \frac{8.85 \times 4\pi \times 10^{-8} \text{ C}}{4\pi \times (0.30) \text{ m}^2}$$

$$= \frac{8.85}{0.09} \times 10^{-8} \text{ C/m}^2 = \textbf{98.33 C/m}^2 \qquad \textbf{... Ans.}$$

Problem 1.13 : *Obtain Coulomb's law from Gauss's law.*

Solution : To obtain *Coulomb's law from Gauss's law,* consider a point charge +q situated at point O in air.

Let us find electric field intensity at point P using Gauss's law. Let d (OP) = r.

To find the electric intensity at point P, imagine a Gaussian surface S in the form of concentric sphere of radius r (Refer Fig. 1.27).

The electric intensity is radially outward and it has the same magnitude at all points on the Gaussian surface. The angle between $\vec{E}$ and normal to any surface element of area dS on the Gaussian surface is zero. Flux passing through small surface element d$\vec{S}$ is

$$d\phi = \vec{E} \cdot d\vec{S} = E \, dS \cos\theta$$

But

$$\theta = 0$$

$$d\phi = E \, dS$$

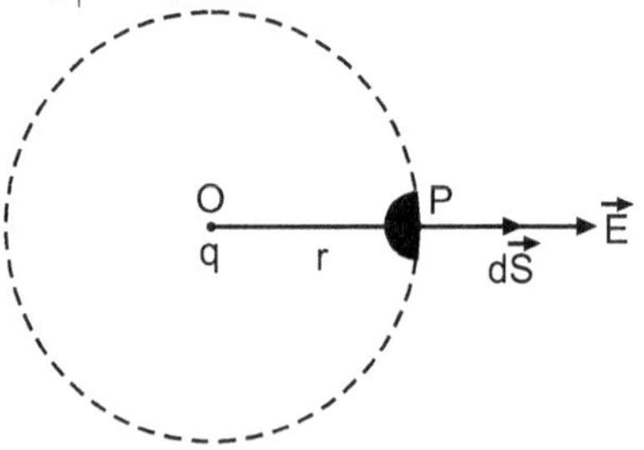

Fig. 1.27 : A spherical Gaussian surface of radius r surrounding a point charge

Total flux through Gaussian surface is

$$\phi = \int d\phi = \oint E \, dS = E \oint dS = E \, 4\pi r^2$$

According to Gauss's theorem, $\phi = \dfrac{q_{enc}}{\varepsilon_0}$

where q_{enc} is the total charge enclosed by Gaussian surface.

$$E \, 4\pi r^2 \;=\; \frac{q}{\varepsilon_0} \qquad\qquad (\because q_{enc} = q)$$

$$E \;=\; \frac{1}{4\pi\varepsilon_0} \, \frac{q}{r^2}$$

This gives the magnitude of the electric intensity $\bar{E}$ at point P, situated at a distance r from an isolated point charge q.

The electric field vector is directed away from q. Now if we put a second point charge q_0 at the point at which E is calculated, it will experience the Coulomb's force $\bar{F}$. The magnitude of the force is given by

$$F \;=\; q_0 E \;=\; \frac{1}{4\pi\varepsilon_0} \frac{qq_0}{r^2}$$

This is precisely Coulomb's law. Thus, we have deduced Coulomb's law from Gauss's law.

Problem 1.14 : *A conductor having a charge density 120 $\mu C/m^2$ is kept in air. Find the magnitude of electric intensity at a point near the conductor.* **(April 2015)**

Solution : Given : Surface charge density r $= 120 \times 10^{-6}$ C/m^2.

To calculate : Magnitude of electric intensity (E).

Formula : $\quad E \;=\; \dfrac{\sigma}{\varepsilon_0} \;=\; \dfrac{120 \times 10^{-6}}{8.85 \times 10^{-12}} \;=\; \dfrac{120}{8.85} \times 10^{6}$

$$E \;=\; \textbf{13.56 N/C} \qquad\qquad \textbf{... Ans.}$$

Problem 1.15 : *Three charges q, −2q and 4q are placed at the corners of an equilateral triangle having length of each side 1.0 m. Compute the potential energy of the structure. (Assume q = 1 $\times 10^{-6}$ coulomb)*

Solution :

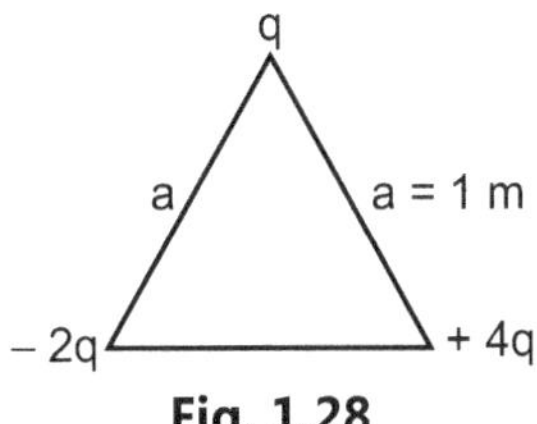

Fig. 1.28

The energy of a pair of charge q_1 and q_2 separated by distance r_{12} is given by, $\dfrac{1}{4\pi\varepsilon_0} \dfrac{q_1 q_2}{r_{12}}$

The total energy of the system of three charges is the algebraic sum of the energies of all pairs of charges.

$$U = \frac{1}{4\pi\varepsilon_0}\left[\frac{q\,(-2q)}{a} + \frac{q\,(4q)}{a} + \frac{(-2q)\,(4q)}{a}\right]$$

$$U = \frac{1}{4\pi\varepsilon_0}\frac{q^2}{a}(-2 + 4 - 8)$$

But, $\dfrac{1}{4\pi\varepsilon_0} = 9 \times 10^9$

$$q = 1 \times 10^{-6}\,C$$

$$a = 1.0 \text{ meter}$$

$$\therefore \quad U = -\frac{9 \times 10^9 \times (1 \times 10^6)^2 \times 6}{1} = -54 \times 10^{-3}\,\textbf{joule} \quad \textbf{... Ans.}$$

Think Over It

1. When you double the charge on both particles in a pair, what effect does this have on the force between them ?
2. In rubbing of glass rod with silk, why is the charge usually transferred by electrons rather than protons ?

Summary

- **Statement of Coulomb's law :** It states that the force of attraction or repulsion between two point charges at rest is directly proportional to the product of magnitude of the charges and inversely proportional to the square of distance between them and is along the line joining them.

  If two point charges q_1 and q_2 are separated by a distance r in some medium, then magnitude of force between them in SI units is

  $$F = \frac{1}{4\pi\varepsilon_0}\frac{q_1\,q_2}{r^2}$$

- **Superposition principle :** It states that all the charges when placed near each other, behave independent of each other and the net force on any particular charge is the vector sum of the forces acting on that charges due to each of the other charge.

- **Electric field intensity ($\vec{E}$) :** The intensity of an electric field at any point in an electric field is defined as the force acting on a unit positive charge placed at that point. **(April 10)**

  The SI unit of electric intensity is N/C. However, an equivalent unit for the electric intensity is volt/meter.

- Electric intensity at a point due to a point charge (q) at a distance r from it is $E = \dfrac{1}{4\pi\varepsilon_o}\dfrac{q}{r^2}$

- **Electric flux (ϕ_E)** : The number of electrical lines of force passing normally through a given surface area drawn in an electric field is called electric flux through that area.

  Mathematically, it is calculated as $d\phi_E = \vec{E}\cdot\vec{dS}$

  S.I. unit of electric flux is $N\cdot m^2/C$.

- **Gauss's theorem in electrostatics** : It states that the total electric flux through any closed surface is equal to q_{enc}/ε_o, where q_{enc} is the total charge inside the closed surface.

  Mathematically, it can be expressed as $\oint \vec{E}\cdot\vec{dS} = \dfrac{q_{enc}}{\varepsilon_o}$

- **Electric Potential:** The electric potential at any point in an electric field is defined as the work done in moving a unit positive charge from infinity to that point against the direction of electric field.

- **Energy of system of charges:** The total work in making an assembly of charges by bringing them from infinity to their respective places is known as the potential energy of system of charges.

Exercises

(A) Multiple Choice Questions :

1. According to Coulomb's law in electrostatics, the magnitude of force between two charges separated by a distance 'r' is proportional to ......

   (a) $\dfrac{1}{r}$

   (b) $\dfrac{1}{r^2}$

   (c) $\dfrac{1}{r^3}$

   (d) r^2

2. The S.I. unit of electric field intensity (E) is ......

   (a) N/C

   (b) N-C

   (c) J/C

   (d) dyne/C

3. The electric intensity at a point situated inside the uniformly charged conducting sphere at a distance r from the centre of sphere is ......

   (a) $E = \dfrac{1}{4\pi\varepsilon_0}\dfrac{q}{r^2}$

   (b) $E = \dfrac{1}{4\pi\varepsilon_0}\dfrac{q}{r}$

   (c) $E = 0$

   (d) $E = \infty$

4. The S.I. unit of electric potential is volt and 1 volt = ...

(a) $\dfrac{1 \text{ erg}}{1 \text{ coulomb}}$

(b) 1 joule - coulomb

(c) $\dfrac{1 \text{ joule}}{1 \text{ coulomb}}$

(d) none of these

5. The potential energy of system of two like charges $+q$ and $+q'$, separated by a distance r is always ......

(a) positive

(b) negative

(c) 0

(d) may be positive or negative

6. S.I. unit of electric flux is ......

(a) $N/m^2 C$

(b) $N \cdot m^2/C$

(c) $N \cdot m^2 \cdot C$

(d) $dyne/cm^2 \cdot C$

7. Electric intensity at a point due to a point charge q at a distance r from it is ......

(a) $E = \dfrac{1}{4\pi\varepsilon_0} \dfrac{qq_0}{r^2}$

(b) $E = \dfrac{1}{4\pi\varepsilon_0} \dfrac{q}{r}$

(c) $E = \dfrac{1}{4\pi\varepsilon_0} \dfrac{q}{r^2}$

(d) $E = \dfrac{1}{4\pi\varepsilon_0} \dfrac{q}{r^3}$

Answers

1. (b)	2. (a)	3. (c)	4. (c)	5. (a)	6. (a)	7. (c)

(B) State True or False Questions :

1. The Coulomb's force between two charges is affected by the presence of other charges.

2. Coulomb's law is applicable in case of point charges at rest only.

3. Gauss's law is not obeyed if the charge distribution on the body is not symmetric and uniform.

4. The potential energy of system of two charges of opposite nature, separated by a distance r is always negative.

5. The electric potential at a point due to charge q at a distance r from it is given by $V = \dfrac{1}{4\pi\varepsilon_0} \dfrac{q}{r^2}$.

Answers

1. False	2. True	3. False	4. True	5. False

(C) Short Answer Type Questions :

1. State Coulomb's law in electrostatics.
2. State Coulomb's law. Express it in vector form.
3. State principle of superposition in electrostatics.
4. Define electric intensity at a point in an electric field. Give its SI unit.
5. What do you mean by electric field or electrostatic field ?
6. What is the value of $\dfrac{1}{4\pi\varepsilon_o}$?
7. Define electric flux.
8. State Gauss's theorem in electrostatics.
9. What do you mean by Gaussian surface ?
10. What are advantages of Gauss's law over Coulomb's law ?
11. Give limitations of Gauss's law.
12. Define the term electric potential.
13. Define the term electric potential energy of a system of charges.

(D) Long Answer Type Questions :

1. State Coulomb's law in electrostatics. Discuss its vector form.
2. State principle of superposition in electrostatics and obtain an expression for force on any one charge due to all other charges.
3. Define electric intensity at a point in an electric field and obtain an expression for electric intensity due to a point charge at any point.
4. Explain the concept of electric flux and give its S.I. unit
5. State and prove Gauss's law. What is the advantage of Gauss's law over Coulomb's law ?
6. State Gauss's law. Give the limitations of Gauss's law.
7. Obtain an expression for the electric intensity near the surface of metallic conductor. (Use Gauss's law).
8. Using Gauss's theorem, obtain an expression for electric intensity at any point due to uniformly charged non-conducting sphere.
9. Using Gauss's theorem, obtain an expression for the electric intensity at any point due to a line charge.
10. Define the term potential energy of a system of charges and obtain an expression for energy of system of charges
11. Define electric intensity at a point in an electric field and obtain an expression for electric intensity at any point due to group of point charges.

(E) Unsolved Problems :

1. Four charges q , –q, q and –q are placed at the corners of square of side a = 1.0 m as shown in Fig. 1.29. Find the potential energy of the system of charges. (Assume q = 1 µC)

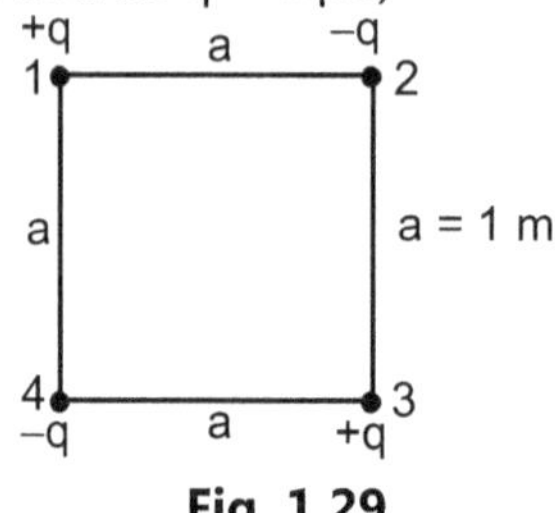

Fig. 1.29

(**Ans.** –23.27 × 10^{-3} joule)

2. The electric field at a point due to a point charge 25 cm away is 10 N/C. What is the magnitude of charge ? (**Ans.** 6.94 × 10^{-11} C)

3. Calculate the electric field intensity required just to support a water droplet of mass 10^{-5} kg and having a charge of 1.6 × 10^{-13} coulomb.
   (**Hint** : Use electric force = gravitational force i.e. qE = mg)

   (**Ans.** 6.125 × 10^{8} volt/meter)

4. Point charges 1 µC, –2 µC, 3 µC and –2 µC are kept at four corners of a square having each side of length 1 cm. Find the net force acting on the charge of 1 µC. (**Ans.** F = 119 N)

5. A charge of 9 × 10^{-9} C is placed inside a cube. Calculate the electric flux linked with the cube. (**Ans.** 1 × 10^{3} Nm²/C)

6. An electric flux of 6.5 × 10^{3} is linked with a sphere due to some charge placed inside the sphere. Calculate the magnitude of charge inside the sphere. (**Ans.** 5.75 × 10^{-8} C)

7. Three point charges each having charge +q are kept on the circumference of a circle in such a way that they make an equilateral triangle. Calculate the electric intensity at the centre of the circle.

   (**Ans.** zero)

8. A charge +q is distributed uniformly throughout a conducting spherical shell of radius 'a'. Calculate electric intensity at a point inside and outside the spherical shell. (Use Gauss's law).

(**Ans.** (i) E at an exterior point is $\dfrac{1}{4\pi\varepsilon_o} \dfrac{q}{r^2}$, (ii) At an internal point, E = 0)

Chapter **2**...

Dielectrics

Contents ...

The energy used in a camera's flash unit is stored in a capacitor, which consists of two closely spaced conductors that carry opposite charges. If charge q is doubled, the stored energy ($W = q^2/2C$) increases by a factor of 4. If value of charge is too great, the electric field magnitude inside the capacitor will exceed the dielectric strength of material and breakdown occurs. This puts practical limit on amount of energy that can be stored.

Flashing of Camera

Rev Dr **William Whewell** DD FRS FGS HFRSE (24 May 1794 – 6 March 1866) was an English polymath, scientist, Anglican priest, philosopher, theologian, and historian of science. He was Master of Trinity College, Cambridge. In his time as a student there, he achieved distinction in both poetry and mathematics.

One of Whewell's greatest gifts to science was his wordsmithing. He often corresponded with many in his field and helped them come up with new terms for their discoveries. Whewell contributed the terms scientist, physicist, linguistics, consilience, catastrophism, uniformitarianism, and astigmatism amongst others; Whewell suggested the terms electrode, ion, dielectric, anode, and cathode to Michael Faraday.

Whewell died in Cambridge in 1866 as a result of a fall from his horse.

2.1 Introduction to Dielectric Material

- In chapter 1, we have studied electric field produced by a given charge distribution in air or vacuum. In the present chapter, we shall study electric field in the presence of dielectric or insulating materials like glass, mica, water, etc. We know about conductors that these are the substances which contain large number of free electrons. When a conductor is placed in an electric field, the electrons move against the direction of field and get distributed in such a way that net electric field inside it is zero. In insulating or dielectric materials, by contrast, positive or negative charges in each atom or molecule are bound together by a strong force and hence these charges are not free to move from one place to another. *A dielectric is an electrically non-conducting material in which there are no free electrons to move. Air, mica, glass, water etc. are examples of dielectrics.* An ideal dielectric is one in which there are no free charges. No current results when an insulating or dielectric material is placed in an electric field. When such a material is kept in an external electric field, the charges undergo minute displacement, giving rise to an electric displacement which is proportional to the applied electric field.

- The charges which are displaced in dielectric material due to an external electric field are called bound charges. When the entire positive charge in the dielectric is displaced with respect to negative charge, the dielectric is said to be polarized. In the present chapter, we begin our study with dipole moment, field and potential produced at a point due to dipole. We then study electric polarization of dielectric material, Gauss's law in dielectrics, etc.

2.2 Electric Dipole

2.2.1 Electric Dipole

- *Two equal and opposite point charges separated by a small distance form an electric dipole.* An electric dipole, consisting of two charges –q and +q separated by a distance 'd' is shown in Fig. 2.1.

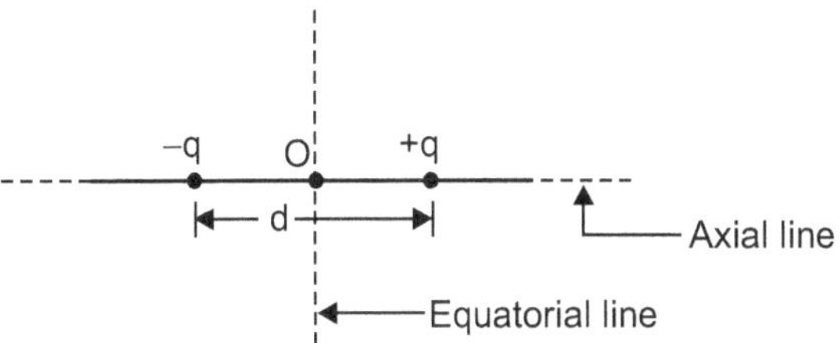

Fig. 2.1 : Electric dipole

- A line joining the two charges is called axial line and a line passing through the centre 'O' of the dipole and perpendicular to the axial line is called equatorial line. The mid point O defines the position of the dipole.

2.2.2 Electric Dipole Moment

- *The product of the magnitude of either charge and the distance between the charges is called electric dipole moment.* It is denoted by $\vec{p}$. The dipole moment is a vector quantity having direction along the line joining the two charges from negative to positive charge.

- Thus, if a dipole consists of two charges $-q$ and $+q$ separated by a distance 'd', the dipole moment is $\vec{p} = q\vec{d}$, where $\vec{d}$ is the vector joining the negative charge to the positive charge.
- S.I. unit of dipole moment is coulomb-meter and C.G.S. (e.s.u.) unit is stat coulomb-cm.

2.3 Electric Potential and Intensity at any Point due to a Dipole

2.3.1 Electric Potential at any Point due to a Dipole

(April 17, 10; Oct. 15)

- Consider an electric dipole consisting of charges $-q$ and $+q$ situated at points A and B respectively. Let the two charges be separated by a small distance d = AB. (Refer Fig. 2.2).
- Let us calculate the potential due to the dipole at point P which is at a distance 'r' from the centre O of the dipole. Since d is small distance, let $r \gg d$.

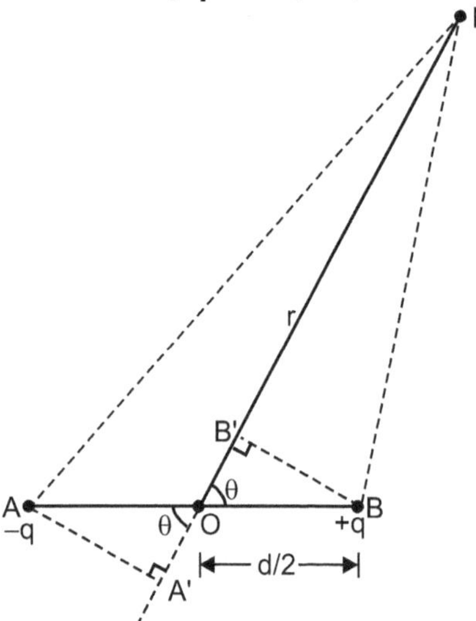

Fig. 2.2 : Potential due to a dipole

The potential at point P due to charge $-q$ is

$$V_1 = -\frac{1}{4\pi\varepsilon_o}\frac{q}{AP}$$

The potential at point P due to charge q is

$$V_2 = \frac{1}{4\pi\varepsilon_o}\frac{q}{BP}$$

The net potential at point P due to dipole is

$$V = V_1 + V_2$$

$$V = \frac{q}{4\pi\varepsilon_o}\left(\frac{1}{BP} - \frac{1}{AP}\right) \qquad \text{... (2.1)}$$

In Fig. 2.2, let $\angle POB = \theta$ and the line BB' (drawn from B to B') is perpendicular to OP.

$$BP \approx B'P = OP - OB'$$

But $\quad OP = r$ and $OB' = \dfrac{d}{2}\cos\theta \qquad\qquad \left(\because \cos\theta = \dfrac{OB'}{d/2}\right)$

$$\therefore \qquad BP = r - \dfrac{d}{2}\cos\theta \qquad\qquad \dots (2.2)$$

Similarly, we have $AP \approx A'P = A'O + OP$

$$AP = r + \dfrac{d}{2}\cos\theta \qquad\qquad \dots (2.3)$$

Substituting equations (2.2) and (2.3) in equation (2.1), we get

$$V = \frac{q}{4\pi\varepsilon_o}\left(\frac{1}{r - \dfrac{d}{2}\cos\theta} - \frac{1}{r + \dfrac{d}{2}\cos\theta}\right) = \frac{q}{4\pi\varepsilon_o}\left[\frac{\left(r + \dfrac{d}{2}\cos\theta\right) - \left(r - \dfrac{d}{2}\cos\theta\right)}{r^2 - \dfrac{d^2}{4}\cos^2\theta}\right]$$

Since $r \gg d$, the term $\dfrac{d^2}{4}\cos^2\theta$ is negligibly small as compared to r^2 and hence it can be neglected.

$$\therefore \qquad V = \frac{q}{4\pi\varepsilon_o}\frac{d\cos\theta}{r^2}$$

But $\qquad qd = p$

$$\therefore \qquad \boxed{V = \frac{1}{4\pi\varepsilon_o}\frac{p\cos\theta}{r^2}} \qquad\qquad \dots (2.4)$$

The above equation gives the potential at a distance r, due to the electric dipole.

We know that, the potential at a distance r from a point charge q is

$$V = \frac{1}{4\pi\varepsilon_o}\frac{q}{r}$$

Thus, in the case of point charge, the potential is inversely proportional to r and is independent of direction; but in the case of dipole, the potential due to it is inversely proportional to r^2 and depends also on the direction as shown by the term $\cos\theta$ (equation 2.4).

Different Cases :

(a) Electric potential at any point on the axial line : If the point P is on the axis of dipole ($\theta = 0°$) and at a distance r from the centre of dipole, the potential at this point is

$$V = \frac{1}{4\pi\varepsilon_o}\frac{p}{r^2}$$

$(\because \cos 0 = 1)$

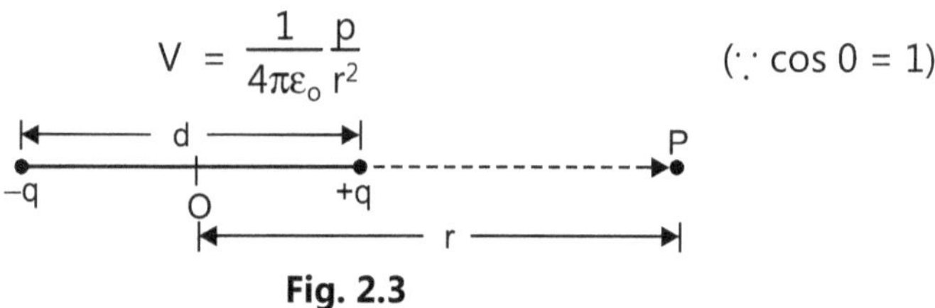

Fig. 2.3

(b) Electric potential at any point on the perpendicular bisector of the dipole (or on the equatorial line) : If the point P is on the perpendicular bisector of the dipole ($\theta = \pi/2$) and at a distance r from the centre of dipole (shown in Fig. 2.4), then the potential at this point is

$$V = \frac{1}{4\pi\varepsilon_o}\frac{p\cos\pi/2}{r^2}$$

But $\cos\pi/2 = 0$

$\therefore$ $V = 0$

Thus, potential at every point on the equatorial line of dipole is zero.

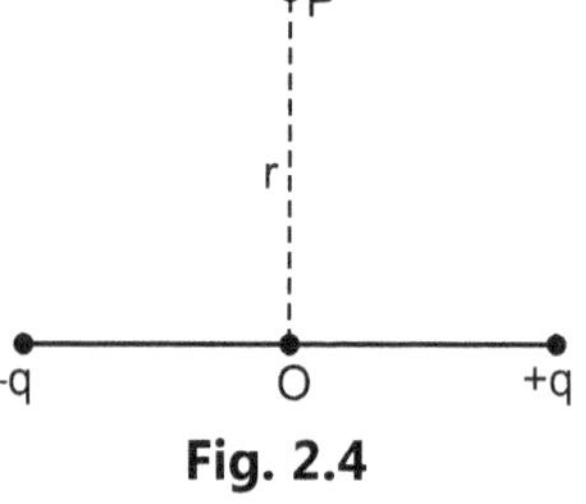

Fig. 2.4

2.3.2 Electric Intensity at any Point due to a Dipole

(April 11)

- Consider an electric dipole consisting of charges $-q$ and $+q$ separated by a small distance d as shown in Fig. 2.5.

- We have to calculate the electric intensity at a point P, which is at a distance r from the centre 'O' of the dipole.

- From the expression of potential due to a dipole (equation 2.4), it is clear that the potential at any point depends on r and θ. In Fig. 2.5, while moving from P to P_1, r changes, but θ remains same and in going from P to P_2, angle θ (i.e. direction) changes but r remains almost constant. The displacement PP_1 is along OP and PP_2 is perpendicular to OP, so PP_1 is called radial direction and PP_2 is called transverse direction.

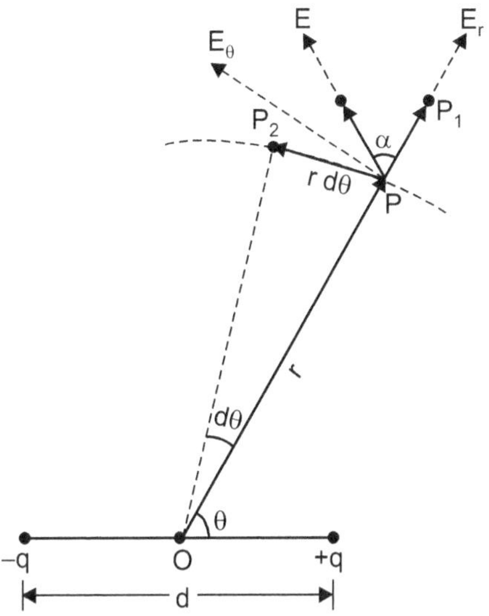

Fig. 2.5 : Electric intensity at a point P due to a dipole

In going from P to P_1, let the change in distance be dr i.e. PP_1 = dr and in going from P to P_2, let the change in angle be dθ. From Fig. 2.5, distance PP_2 = rdθ.

Thus the component of electric field intensity at a point P in the radial direction PP_1 is $E_r = -\dfrac{dV}{PP_1} = -\dfrac{\partial V}{\partial r}$

The symbol ∂ is used because while differentiating w.r.t. r, θ remains constant.

Similarly, the component of electric field intensity at a point P in the transverse direction PP_2 is

$$E_\theta = -\frac{dV}{PP_2} = -\frac{\partial V}{r\partial\theta} \qquad (\because PP_2 = r\,d\theta)$$

The potential at a point P due to electric dipole is

$$V = \frac{1}{4\pi\varepsilon_o}\frac{p\cos\theta}{r^2}$$

$$\therefore \qquad E_r = -\frac{\partial V}{\partial r} = -\frac{\partial}{\partial r}\left(\frac{1}{4\pi\varepsilon_o}\frac{p\cos\theta}{r^2}\right)$$

$$E_r = \frac{1}{4\pi\varepsilon_o}\frac{2p\cos\theta}{r^3} \qquad \left(\because \frac{\partial}{\partial r}\frac{1}{r^2} = -\frac{2}{r^3}\right)$$

and
$$E_\theta = -\frac{\partial V}{r\partial\theta} = -\frac{\partial}{r\partial\theta}\left(\frac{1}{4\pi\varepsilon_o}\frac{p\cos\theta}{r^2}\right) = -\frac{p}{4\pi\varepsilon_o\, r^3}\frac{\partial}{\partial\theta}\cos\theta$$

$$E_\theta = \frac{1}{4\pi\varepsilon_o}\frac{p\sin\theta}{r^3}$$

The resultant electric intensity at a point P is

$$E = \sqrt{E_r^2 + E_\theta^2}$$

$$= \frac{1}{4\pi\varepsilon_o}\sqrt{\left(\frac{2p\cos\theta}{r^3}\right)^2 + \left(\frac{p\sin\theta}{r^3}\right)^2}$$

$$= \frac{1}{4\pi\varepsilon_o r^3}\sqrt{4p^2\cos^2\theta + p^2\sin^2\theta}$$

$$= \frac{p}{4\pi\varepsilon_o r^3}\sqrt{4\cos^2\theta + \sin^2\theta}$$

$$E = \frac{p}{4\pi\varepsilon_o r^3}\sqrt{3\cos^2\theta + 1} \quad (\because \sin^2\theta = 1 - \cos^2\theta) \ ... \ (2.5)$$

Equation (2.5) represents the expression for electric intensity at a point P due to a dipole.

If the resultant intensity E makes an angle α with the direction OP, we have

$$\tan\alpha = \frac{E_\theta}{E_r} = \frac{\dfrac{1}{4\pi\varepsilon_o}\dfrac{p\sin\theta}{r^3}}{\dfrac{1}{4\pi\varepsilon_o}\dfrac{2p\cos\theta}{r^3}}$$

$$\tan\alpha = \frac{1}{2}\tan\theta$$

$$\alpha = \tan^{-1}\left(\frac{1}{2}\tan\theta\right)$$

Different cases :

(a) If the point P is on the axis of dipole ($\theta = 0$) and at a distance r from the dipole, the electric intensity at this point is

$$E = \frac{1}{4\pi\varepsilon_o}\frac{2p}{r^3} \qquad\qquad (\because \theta = 0°)$$

(b) If the point P is on the perpendicular bisector of the dipole $\left(\theta = \dfrac{\pi}{2}\right)$, the electric intensity at this point is

$$E = \frac{1}{4\pi\varepsilon_o}\frac{p}{r^3} \qquad \left(\because \cos\frac{\pi}{2} = 0\right)$$

2.4 Torque on a Dipole Placed in an Electric Field   (April 16, Oct. 16)

- Consider an electric dipole consisting of charge '–q' placed at point A and charge '+q' placed at point B in an uniform electric field $\overrightarrow{E}$ as shown in Fig. 2.6. The mid point of AB is O and the length AB = d.

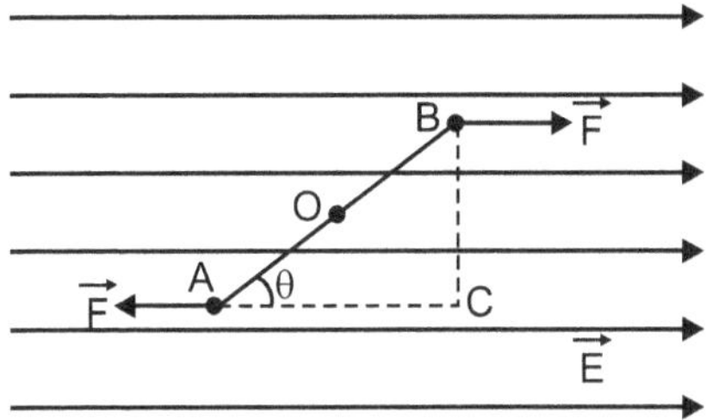

Fig. 2.6 : Electric dipole in a uniform electric field

- Let the axis of the dipole AB make an angle θ with the direction of electric field. The force on the charge q is $\overrightarrow{F_1} = q\overrightarrow{E}$ and the force on the charge –q is $\overrightarrow{F_2} = -q\overrightarrow{E}$. Thus, the resultant force on the dipole from the field is zero. Since these forces do not act along a line, these constitute a couple which exerts a torque τ on the dipole.

$$\text{Torque, } \tau = \begin{pmatrix}\text{Magnitude of}\\ \text{either force}\end{pmatrix} \times \begin{pmatrix}\text{Perpendicular distance}\\ \text{between the two forces}\end{pmatrix}$$

$$\tau = qE \times BC$$

From Fig. 2.6, $\sin\theta = \dfrac{BC}{d}$ or $BC = d\sin\theta$

$\therefore$ $\tau = qE\,d\sin\theta$

But $qd = p$, dipole moment of dipole

$\therefore$ $\tau = pE\sin\theta$

In vector form, the above expression may be expressed as

$$\vec{\tau} = \vec{p} \times \vec{E}$$

The direction of the torque is perpendicular to the plane containing the dipole axis and the electric field. The directions

of $\vec{\tau}$, $\vec{p}$ and $\vec{E}$ are as shown in Fig. 2.7.

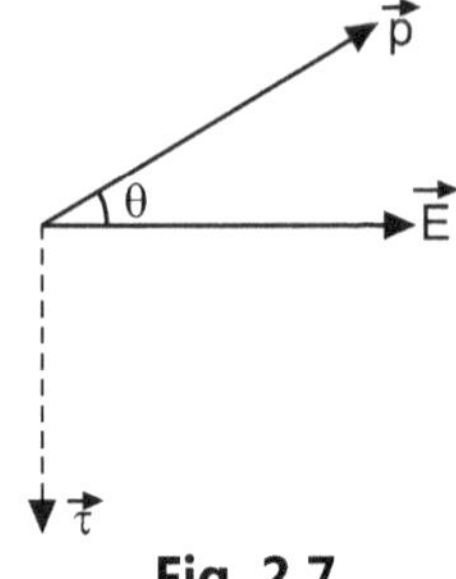

Fig. 2.7

- Thus the tendency of the torque is to bring the dipole along the direction of the field. We have seen that in an uniform field, two charges of dipole experience equal and opposite force, so the net force on the dipole is zero. But in a non-uniform field, the forces on the two charges will not be equal and opposite and as a result, a net force would act on the dipole.

2.5 Polar and Non-Polar Molecules
(April 16; Oct. 16, 15)

Polar molecules :

- *A polar molecule is one in which the centre of gravity of positive charges is separated from a centre of gravity of negative charges by a finite distance of molecular dimensions.* This molecule thus has permanent dipole moment directed from negative to positive centre of gravity of charges. Some common examples of polar molecules are HCl, CO, H_2O, NH_3, C_6H_5Cl, etc.

- In HCl molecule, due to high electron affinity of Cl, the electron of the H atom lies more towards the Cl atom than the H atom. Thus, Cl end of the molecule is negative and H end is positive.

  It is shown in Fig. 2.8 (a). The dipole moment $\vec{p}$ is directed from Cl atom to H atom. Similarly the CO molecule is polar having permanent dipole moment directed from O to the C atom.

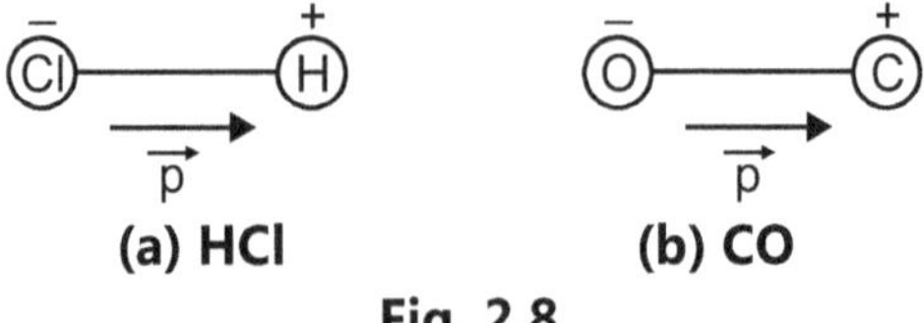

(a) HCl   **(b) CO**

Fig. 2.8

- H_2O molecule is also polar molecule. Its structure is triangular as shown in Fig. 2.9 (a). O atom has 2 unit negative charge and each H atom has 1 unit positive charge. Each H–O bond has dipole moment $\vec{p_1}$ directed from O to the H atom. The resultant dipole moment $\vec{p}$ is directed from O towards the H–H base as shown in Fig. 2.9 (b).

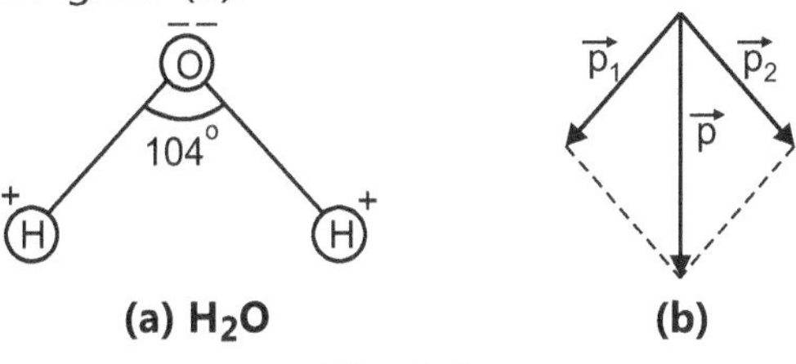

(a) H_2O   **(b)**

Fig. 2.9

Non-polar molecules :

- *A non-polar molecule is the one in which the centre of gravity of the positive charges (protons) coincides with the centre of gravity of negative charges (electrons).* This molecule thus does not have permanent dipole moment. The molecules having a symmetrical structure are non-polar. e.g. H_2, N_2, O_2, CO_2, CH_4, CCl_4, etc.

- Let us consider CO_2 molecule. It has linear symmetrical structure as shown in Fig. 2.10. The dipole moments $\vec{p_1}$ and $\vec{p_2}$ are equal in magnitude and opposite in direction. So the resultant dipole moment $\vec{p}$ = 0. Thus, CO_2 molecule is non-polar.

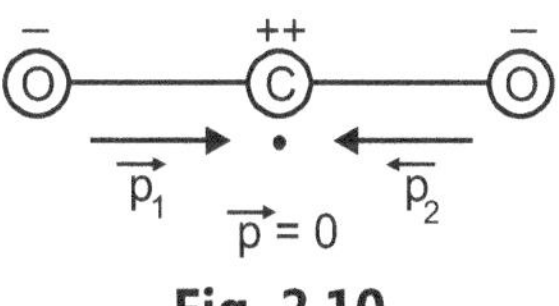

Fig. 2.10

Effect of External Electric Field on Polar and Non-Polar Molecules:

- When non-polar molecule is placed in an external electric field $\vec{E}$, the electric field causes a force to be exerted on each charged centre. Positive charge centre is pushed along the direction of field and negative centre opposite to the direction of the field. Thus, the molecule or atom acquires an induced electric dipole moment along the direction of field as shown in Fig. 2.11 (b).

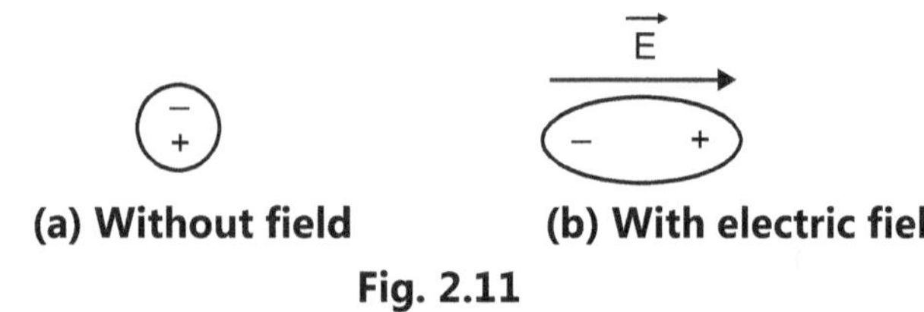

(a) Without field **(b) With electric field**

Fig. 2.11

- In a dielectric material of polar molecules, the dipole moments of individual atoms or molecules are randomly oriented. When such a molecule is placed in an external electric field, there is a tendency for the permanent dipole to align itself along the direction of field. (Refer Fig. 2.12).

(a) Without field **(b) With electric field**

Fig. 2.12

2.6 Electric Polarization of Dielectric Material

Dielectric Materials

- *A dielectric is an electrically non-conducting material in which there are no free electrons to move. Air, mica, glass, water etc. are examples of dielectrics.* An ideal dielectric is one in which there are no free charges. No current results when an insulating or dielectric material is placed in an electric field. The electrons remain firmly locked to their atoms or molecules. Instead of moving charges through material, an external electric field produces a slight rearrangement of electric charges within the atoms. Depending upon type of molecules, the dielectrics are of two types viz. polar and non-polar. Polar dielectric is made up of polar molecules and non-polar dielectric is made up of non-polar molecules. We can use two terms dielectric material or insulating material interchangeably. However, usually if the main purpose of the non-conducting material is just to provide insulation, then it is referred to as insulating material; but **if it is employed for charge storage, then its name dielectric remains.**

2.6.1 Electric Polarization of Dielectric Material

(April 11)

- Let us consider first isotropic non-polar dielectric material placed in an external electric field. Without external electric field, the molecules of the dielectric do not possess individual dipole moments. In an external electric field, the positive and negative charge centres of non-polar molecules, or atoms, move due to electrostatic forces in opposite directions. In each molecule or atom, the centres of gravity of the two charges are separated from each other within molecular distance. Thus each molecule acquires an induced electric dipole moment in the direction of applied field as shown in Fig. 2.13 (b).

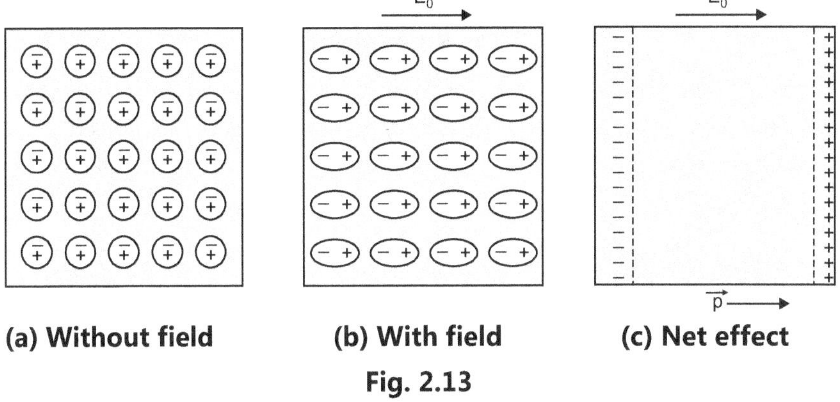

(a) Without field   **(b) With field**   **(c) Net effect**

Fig. 2.13

- The overall effect from macroscopic point of view is a displacement of the net positive charge in the dielectric relative to the negative charge [as shown in Fig. 2.13 (c)], and the dielectric is said to be polarized.

- In a dielectric material of polar molecules, each molecule possesses net dipole moment. The dipole moments of molecules are randomly oriented giving net dipole moment of the material zero as shown in Fig. 2.14 (a). When such a dielectric is placed in an external electric field, the dipole moments of the molecules tend to be aligned in the direction of the field as shown in Fig. 2.14 (b). The alignment is, however, incomplete due to the thermal motion of the molecules. The alignment becomes more and more perfect when electric field is increased and temperature is decreased. Hence net dipole moment is produced along the field direction in the dielectric as shown in Fig. 2.14 (c).

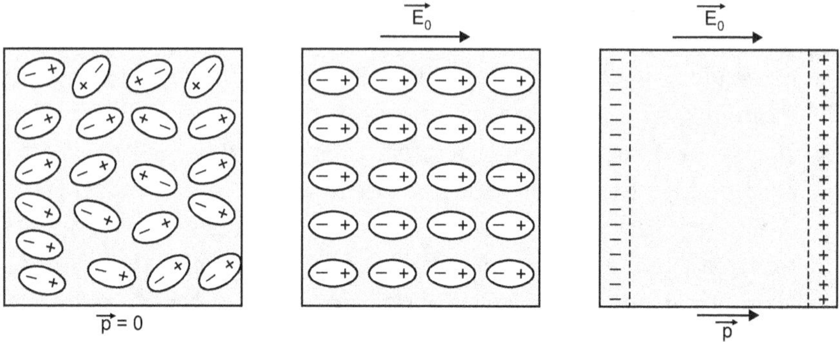

(a) Without field   (b) With field   (c) Net effect

Fig. 2.14 : Polar dielectric

- *Thus, when a dielectric (polar or non-polar) is placed in an electric field, a net dipole moment in the direction of field is produced. This phenomenon is known as electric polarization of dielectric material.*

- Though the dielectric is polarized in an external field, it is electrically neutral.

2.6.2 Electric Polarization Vector $(\vec{P})$ (April 16)

- When a dielectric is placed in an electric field, the net dipole moment is produced in the material along the direction of field and the dielectric is said to be polarized.

- *The net electric dipole moment per unit volume of the dielectric material is known as the electric polarization vector or polarization density.* It is denoted by $\vec{P}$. Since the electric dipole moment is a vector, the electric polarization is also a vector quantity.

- Let us consider a slab of homogeneous isotropic dielectric of cross-sectional area A and thickness t. Let it be placed perpendicular to uniform electric field $\vec{E_0}$. The slab is polarized as shown in Fig. 2.15.

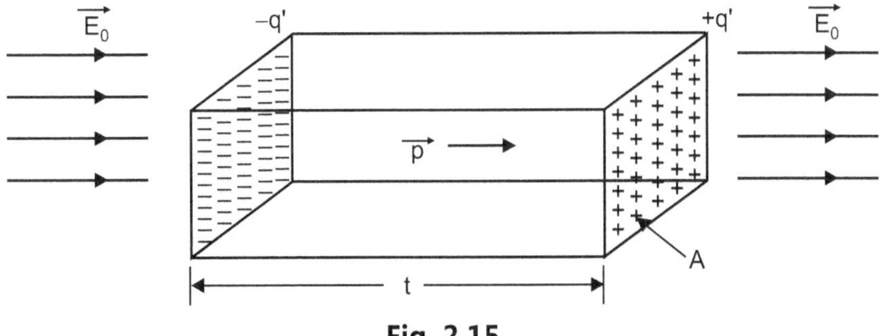

Fig. 2.15

- Let $-q'$ and q' be the net induced charges on the two opposite faces of the slab. These charges are called bound charges because these are bound to the dielectric material.

- The dipole moment of the slab as a whole is

$$p = q't$$

and is directed from negative induced charge $(-q')$ to positive induced charge (q').

The volume of the slab is $V = At$.

The electric polarization is $\vec{P} = \dfrac{\vec{p}}{V}$

The magnitude of $\vec{P}$ is $P = \dfrac{q't}{At} = \dfrac{q'}{A}$

Thus, magnitude of polarization P is net induced charge per unit area.

But, $\dfrac{q'}{A}$ is the surface charge density σ'. Thus, $P = \sigma'$

Thus, for a homogeneous isotropic dielectric, *the electric polarization P is equal to surface charge density of induced charges.*

S.I. unit of P is coulomb / m^2.

The direction of $\vec{P}$ is from the negative induced charge $-q'$ to the positive induced charge $+q'$ i.e. it is directed along the direction of external field. It is due to bound charges or induced charges. In vacuum, it is zero.

2.6.3 Electric Field in Dielectric

- When a dielectric is placed in an electric field, there are two possibilities :

    (i) In dielectric of polar molecules, the permanent dipole moments of molecules are aligned along the field direction and the dielectric is polarized.

    (ii) In dielectric of non-polar molecules, dipole moment is induced in each atom or molecule along the field direction and the dielectric is polarized.

- In both the cases, the surface, which is on positive side of electric field, acquires net negative charge and opposite face acquires net positive charge as shown in Fig. 2.16. The net charge in the interior of slab is zero because the positive side of one polarized molecule is adjacent to the negative side of its neighbour as shown in Fig. 2.13 (b).

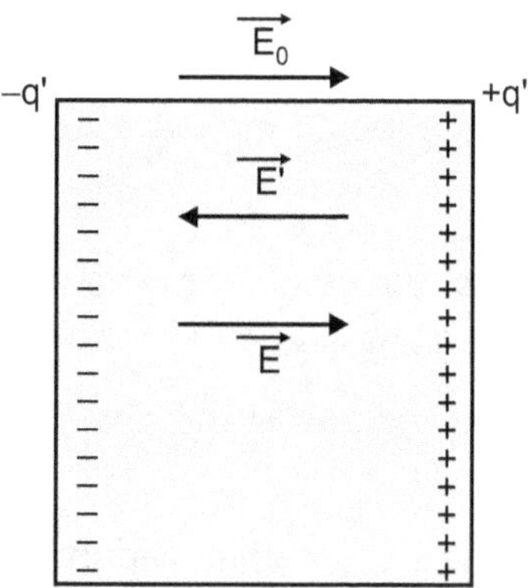

Fig. 2.16

- The polarization charges q' and −q' induced on the two faces of the slab produce their own electric field E', directed from q' to −q'. It is opposite to external field $\overrightarrow{E_o}$. Hence, resultant field within the dielectric is

$$\overrightarrow{E} = \overrightarrow{E_o} - \overrightarrow{E'}$$

$\overrightarrow{E}$ is directed along the direction of external electric field. The field outside the dielectric is still $\overrightarrow{E_o}$. Thus, we conclude that, when a dielectric is placed in an external electric field, the field within the dielectric is weakened (but not reduced to zero).

- The ratio of field outside the dielectric to within the dielectric is greater than one. It is denoted by k and called **dielectric constant.**

Therefore, $\dfrac{E_o}{E} = k$

Thus, the electric field within the dielectric is reduced by a factor k.

2.7 Gauss's Law in Dielectrics

(April 15; Oct. 15, 11, 10)

- The Gauss's law in free space states that *the electric flux passing through a closed surface is equal to $\dfrac{1}{\varepsilon_o}$ times the net charge enclosed by the surface.*

- Consider a parallel-plate capacitor with plate area A and separation between the plates 'd' and having vacuum between the plates. Let +q and −q are the charges on plates as shown in Fig. 2.17 (a). Due to these charges, $\vec{E_o}$ is the uniform electric field between the plates directed from positive plate to negative plate.

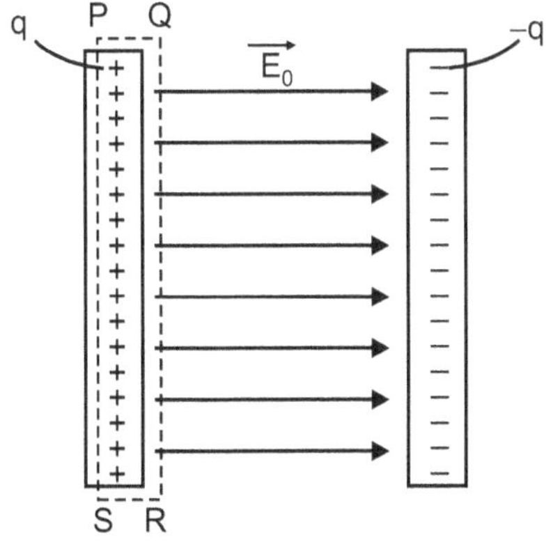

(a) Without dielectric

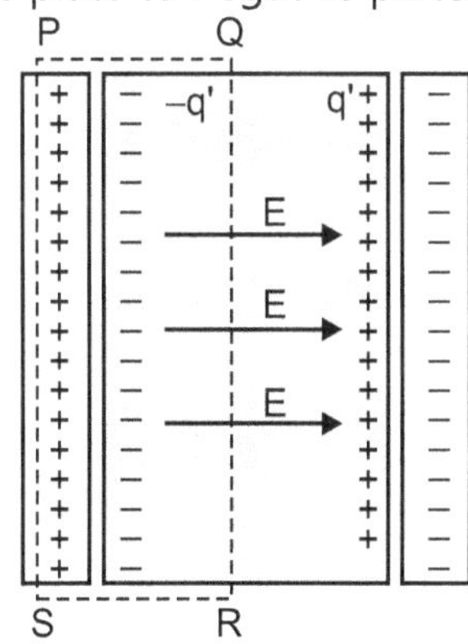

(b) With dielectric

Fig. 2.17

Let us consider Gaussian surface PQRS. According to Gauss's law, the electric flux passing through the closed surface is

$$\oint \vec{E_o} \cdot d\vec{A} = \frac{q}{\varepsilon_o}$$

The electric flux will pass through the surface QR only. As the field is normal to the surface, the angle between normal to the area and electric field is zero.

Here $\overrightarrow{dA}$ is the small vector area on the surface QR. But

$$\oint \overrightarrow{E_o} \cdot \overrightarrow{dA} = E_o A.$$

$$\therefore \qquad E_o A = \frac{q}{\varepsilon_o}$$

$$\text{or} \qquad E_o = \frac{q}{A\varepsilon_o} \qquad\qquad \dots (2.6)$$

Let us now suppose that a dielectric material of dielectric constant is filled completely between the plates [Fig. 2.17 (b)]. A negative charge $-q'$ is induced on the surface which is on the side of positive plate of capacitor and q' is induced on the surface on the side of negative plate. Again consider a Gaussian surface PQRS. The net charge enclosed by the Gaussian surface is $q - q'$. Let the electric field within the dielectric be $\overrightarrow{E}$.

Applying Gauss's law, we get

$$\oint \overrightarrow{E} \cdot \overrightarrow{dA} = \frac{q - q'}{\varepsilon_o} \qquad\qquad \dots (2.7)$$

$$EA = \frac{q - q'}{\varepsilon_o} \quad \text{or} \quad E = \frac{q}{A\varepsilon_o} - \frac{q'}{A\varepsilon_o} \qquad\qquad \dots (2.8)$$

But we have $\qquad \dfrac{E_o}{E} = k \ \text{ or } E = \dfrac{E_o}{k}$

Using equation (2.6) in the above equation, we get

$$E = \frac{q}{k\varepsilon_o A} \qquad\qquad \dots (2.9)$$

Comparing equations (2.8) and (2.9), we get

$$\frac{q}{k\varepsilon_o A} = \frac{q}{A\varepsilon_o} - \frac{q'}{A\varepsilon_o} \qquad\qquad \dots (2.10)$$

Therefore, net induced charge on each surface of dielectric is

$$q' = q\left(1 - \frac{1}{k}\right)$$

Since k > 1, the induced charge q' is always less than the free charge q on the plates of capacitor. q' is equal to zero if no dielectric is present.

Above equation may be written as $q - q' = \dfrac{q}{k}$ ... (2.11)

Using equation (2.11) in equation (2.7), we get

$$\oint \vec{E} \cdot d\vec{A} = \dfrac{q}{k\varepsilon_o} \qquad \ldots (2.12)$$

i.e. $\oint \vec{E} \cdot d\vec{A} = \dfrac{q}{\varepsilon}$ where $k = \dfrac{\varepsilon}{\varepsilon_o}$ or $\varepsilon = k\varepsilon_o$

or $\oint \varepsilon \vec{E} \cdot d\vec{A} = q$ ... (2.13)

$$\oint \vec{D} \cdot d\vec{A} = q \qquad \ldots (2.14)$$

where $\vec{D} = \varepsilon \vec{E} = k\varepsilon_o \vec{E}$, called *electric displacement vector.*

Equation (2.14) represents Gauss's law in dielectrics. It states that *the surface integral of electric displacement vector $\vec{D}$ over a closed surface is equal to free charges enclosed within the surface.*

Electric displacement D depends on only free charges. The induced charge q' is taken into account by the introduction on left side of equation (2.14).

2.8 Electric Vectors and Relation Between Them
(Relation between $\vec{E}$, $\vec{D}$ and $\vec{P}$)   (April 17)

For a parallel-plate capacitor containing a dielectric material of dielectric constant k, we have obtained equation (2.10) as

$$\dfrac{q}{k\varepsilon_o A} = \dfrac{q}{A\varepsilon_o} - \dfrac{q'}{A\varepsilon_o}$$

$\therefore$ $\dfrac{q}{A\varepsilon_o} = \dfrac{q}{k\varepsilon_o A} + \dfrac{q'}{A\varepsilon_o}$

or $\dfrac{q}{A} = \varepsilon_o\left(\dfrac{q}{k\varepsilon_o A}\right) + \dfrac{q'}{A}$

The quantity in the bracket of above equation represents the electric intensity in the dielectric i.e. $E = \dfrac{q}{k\varepsilon_o A}$ (See equation 2.9). The last term in the above equation is induced charge per unit area (q'/A). It is known as electric polarization i.e. $P = q'/A$.

Therefore, above equation can be written as

$$\frac{q}{A} = \varepsilon_o E + P$$

The quantity on the left hand side of above equation is called the electric displacement (D).

$$\therefore \qquad D = \varepsilon_o E + P \text{ where, } D = \frac{q}{A}$$

Since $\vec{E}$ and $\vec{P}$ are vectors, $\vec{D}$ is also a vector.

Therefore, the relation between three vectors $\vec{D}$, $\vec{E}$ and $\vec{P}$ is given by, $\qquad \vec{D} = \varepsilon_o \vec{E} + \vec{P} \qquad\qquad …\ (2.15)$

The three vectors are shown in Fig. 2.18.

We must note the following :

1. $\vec{D}$ is connected with the free charge or true charge only, so it is not altered by the introduction of the dielectric. We can represent the vector field of $\vec{D}$ by lines of $\vec{D}$, just like lines of electric field $\vec{E}$ which is shown in Fig. 2.18.

2. $\vec{P}$ is connected with polarization (bound) charges or induced charges only. It is represented by lines directed from the negative induced charge to the positive induced charge and present within the dielectric.

3. $\vec{E}$ is connected with all charges that are actually present, whether free or bound. As shown in Fig. 2.18, the lines of $\vec{E}$ indicate the presence of both kinds of charges.

Note that the SI unit of $\vec{D}$ and $\vec{P}$ is coulomb/m^2, while SI unit of $\vec{E}$ is newton/coulomb.

The relation between D, P and E is given as

$$D = \varepsilon_o E + P \quad \text{or} \quad P = D - \varepsilon_o E$$

Since $D = k\varepsilon_o E$, we get $P = \varepsilon_o (k - 1) E$ ... (2.16)

This is the relation between electric polarization (polarization density) and electric intensity.

In vector form, $\vec{P} = \varepsilon_o (k - 1) \vec{E}$

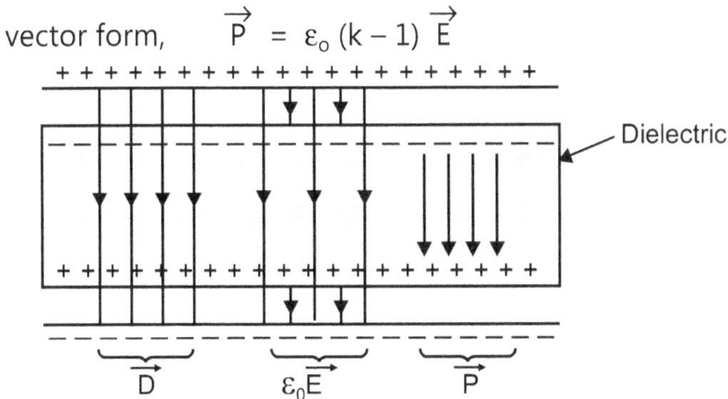

Fig. 2.18

Solved Problems

Problem 2.1 : *An electric dipole consisting of two opposite charges each of magnitude 2.00 μC is separated by a distance of 2.0 cm. The dipole is placed in an external field of intensity 1.0×10^5 N/C. Calculate the maximum torque on the dipole.*   **(April 17)**

Solution : Given : $q = 2.00 \,\mu C = 2 \times 10^{-6}$ C

$$d = 2.0 \text{ cm} = 2 \times 10^{-2} \text{ m}$$
$$E = 1.0 \times 10^5 \text{ N/C}$$

Formula : Torque, $\tau = pE \sin \theta$

Maximum torque, $\tau = pE$

where p is dipole moment and E is electric intensity of the field.

$\therefore$ $\tau = 2 \times 10^{-6} \times 2 \times 10^{-2} \times 1.0 \times 10^5$ $(\because p = qd)$

$$\boldsymbol{\tau = 4 \times 10^{-3} \text{ Nm}}$$ **... Ans.**

Problem 2.2 : *Calculate the potential and electric field due to a dipole of dipole moment 2×10^{-10} C-m at a distance of 1 metre from it (a) on its axis, (b) on its perpendicular bisector.*   **(April 2015)**

Solution : Given : (1) Electric dipole moment, $p = 2 \times 10^{-10}$ C-m.

(2) Distance of the point from position of the dipole, $r = 1$ m.

Formula : (i) Electric potential at any point is

$$V = \frac{1}{4\pi\varepsilon_o} \frac{p\cos\theta}{r^2}$$

If the point is on the axis of dipole then electric potential,

$$V = \frac{1}{4\pi\varepsilon_o} \frac{p}{r^2} \qquad (\because \cos 0 = 1)$$

$$= \frac{9\times 10^9 \times 2\times 10^{-10}}{1\times 1} = 18\times 10^{-1} \text{ V}$$

or $V = \textbf{1.8 V}$ **... Ans.**

Thus, electric potential at a point on the axis of the dipole situated at a distance of 1 metre from the dipole = 1.8 V.

(ii) The electric intensity at any point due to electric dipole is

$$E = \frac{p}{4\pi\varepsilon_o r^3} \sqrt{3\cos^2\theta + 1}$$

If the point is on the axis of dipole then electric intensity,

$$E = \frac{p}{4\pi\varepsilon_o r^3} \times 2 \qquad (\because \cos 0 = 1)$$

$$= \frac{9\times 10^9 \times 2\times 10^{-10} \times 2}{1\times 1\times 1} = \textbf{3.6 V/m} \qquad \textbf{... Ans.}$$

Thus, electric intensity at a point on the axis of the dipole situated at a distance of 1 metre from the dipole = 3.6 V/m.

(iii) If the point is on the perpendicular bisector, then electric potential at any point is $V = 0$ $(\because \cos 90° = 0)$

Thus, electric potential at a point on the perpendicular bisector situated at a distance of 1 metre from the dipole = 0 V.

(iv) The electric intensity at a point on the perpendicular bisector situated at a distance of 1 metre from the dipole is

$$E = \frac{p}{4\pi\varepsilon_o r^3} \qquad (\because \cos 90° = 0)$$

$$= \frac{9\times 10^9 \times 2\times 10^{-10}}{1\times 1\times 1}$$

$$E = \textbf{1.8 volts/meter} \qquad \textbf{... Ans.}$$

Problem 2.3 : *A dielectric slab of thickness 0.6 cm and dielectric constant k = 5 is placed between the parallel plates of plate area 0.01 m^2 and separation 0.015 m. A potential difference of 150 volt is applied with no dielectric present. If the battery is connected and dielectric is inserted, find the three vectors E, D and P in the dielectric.*

Solution : Given : V_o = 150 volts, A = 0.01 m^2, d = 0.015 m, k = 5,

$$\varepsilon_o = 8.85 \times 10^{-12} \; C^2/N\text{-}m^2$$

The electric intensity in air between the two plates is

$$E_o = \frac{V_o}{d} = \frac{150}{0.015} = 1 \times 10^4 \; V/m.$$

When dielectric is inserted, the electric intensity in the dielectric reduces by a factor k.

$$E = \frac{E_o}{k} = \frac{1 \times 10^4}{5} = \mathbf{2 \times 10^3 \; V/m} \qquad \textbf{... Ans.}$$

The electric displacement in the dielectric is given by

$$D = k\varepsilon_o E = 5 \times 8.85 \times 10^{-12} \times 2 \times 10^3$$
$$= \mathbf{8.85 \times 10^{-8} \; C/m^2} \qquad \textbf{... Ans.}$$

The electric polarization in the dielectric is given by

$$P = \varepsilon_o(k - 1) E = 8.85 \times 10^{-12} \times (5 - 1) \times 2 \times 10^3$$
$$= \mathbf{7.08 \times 10^{-9} \; C/m^2} \qquad \textbf{... Ans.}$$

Problem 2.4 : *Two parallel plates have equal and opposite charges. When the space between the two plates is vacuum, the electric intensity is 3 $\times$ 10^6 V/m. When the space is filled with dielectric, the electric intensity becomes 1.5 $\times$ 10^6 V/m. Find the induced charge density on the surface of the dielectric.*

Solution : Electric field without dielectric E_o = 3 $\times$ 10^6 V/m

Electric field with dielectric E = 1.5 $\times$ 10^6 V/m

When dielectric is inserted, the electric intensity in the dielectric reduces by a factor k.

$$E = \frac{E_o}{k}$$

$$\therefore \quad k = \frac{E_o}{E} = \frac{3 \times 10^6}{1.5 \times 10^6} = 2$$

The electric polarization in the dielectric is given by

$$P = \varepsilon_o (k - 1) E = 8.85 \times 10^{-12} \times (2 - 1) \times 1.5 \times 10^6$$

$$= 13.275 \times 10^{-6} \text{ C/m}^2 \qquad \text{... Ans.}$$

Polarization P represents induced surface charge density $\dfrac{q'}{A}$.

Problem 2.5 : *Two point charges in a dielectric medium having k = 5.2 interact with a force of 8.6 $\times 10^{-3}$ N. What could be the force if the charges were in free space ?* **(Oct. 16)**

Solution : Given : (i) Dielectric constant, k = 5.2.

(ii) Force between two charges when situated in dielectric material, $\quad F = 8.6 \times 10^{-3}$ N

When two point charges are situated in free space, force between the two charges is given by,

$$F_{air} = \frac{1}{4\pi\varepsilon_o} \frac{q_1 q_2}{r^2}$$

When two point charges are situated in dielectric medium with dielectric constant k, force between the two charges is given by,

$$F = \frac{1}{4\pi\varepsilon_o k} \frac{q_1 q_2}{r^2}$$

$$\therefore \qquad F_{air} = F \times k = 8.6 \times 10^{-3} \times 5.2 = \textbf{4.4} \times \textbf{10}^{-2} \textbf{ newton... Ans.}$$

Problem 2.6 : *A parallel-plate capacitor of plate area A = 100 cm^2 and separation d = 1.5 cm is charged by a potential of 60 V. Then the battery is disconnected and dielectric slab of thickness b = 0.8 cm and k = 5 is inserted. Calculate three vectors E, D and P in the dielectric.*

Solution : Given : (i) Area of plate, A = 100 cm^2.

(ii) Separation or distance between two plates, d = 1.5 cm.

(iii) Potential V = 60 V.

The electric intensity between two plates without dielectric is,

$$E_o = \frac{V}{d} = \frac{60 \text{ volt}}{1.5 \times 10^{-2} \text{ m}} = 40 \times 10^2 \text{ volt/m}$$

When the dielectric slab of k = 5 is inserted, the electric intensity inside the dielectric becomes,

$$E = \frac{E_o}{k} = \frac{40 \times 10^2}{5} = \textbf{8} \times \textbf{10}^2 \textbf{ volt/m} \qquad \text{... Ans.}$$

The electric displacement vector is given by,

$$D = k \, \varepsilon_0 \, E$$
$$D = 5 \times 8.85 \times 10^{-12} \times 8 \times 10^2$$
$$D = \mathbf{3.54 \times 10^{-8} \ C/m^2} \qquad \textbf{... Ans.}$$

The electric polarization in the dielectric is given by,

$$P = (k - 1) \, \varepsilon_0 \, E$$
$$\therefore \qquad P = (5 - 1) \times 8.85 \times 10^{-12} \times 8 \times 10^2$$
$$= 4 \times 8.85 \times 8 \times 10^{-10} = \mathbf{2.83 \times 10^{-8} \ C/m^2} \qquad \textbf{... Ans.}$$

Problem 2.7 : *An infinitely long wire having a charge 'q' per unit length is surrounded by rubber insulation out to a radius 'a'. Find magnitude of electric displacement vector $\overrightarrow{D}$ at any point.*

Solution : Fig. 2.19 shows a part of long wire having a charge q per unit length and is surrounded by rubber insulation out to a radius 'a'.

To find the electric intensity at a point P situated inside the rubber insulation at a perpendicular distance r from the axis of wire, imagine a cylinder of radius r and length L co-axial to wire. This cylinder (shown by dotted line) is Gaussian surface and the point P is on its surface.

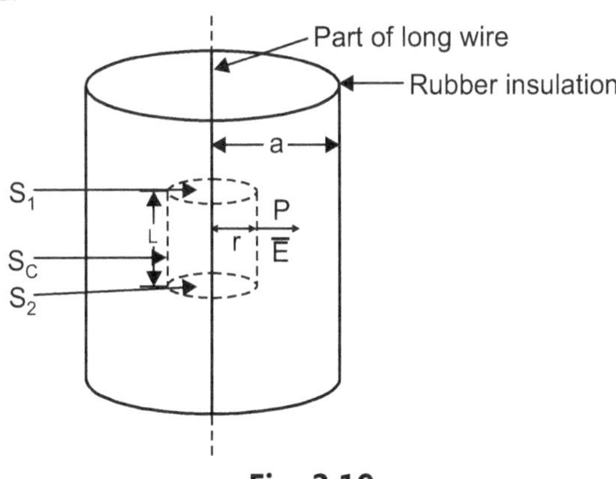

Fig. 2.19

Let $\overrightarrow{D}$ be the electric displacement vector at a point P.

Since there is no electric field perpendicular to the plane faces of the Gaussian surface, flux passing through these faces (S_1 and S_2) is zero.

Flux through the Gaussian surface (curved surface S_C) is given by,

$$\int \vec{D} \cdot d\vec{S} = q_{enc.}$$

But, $q_{enc.} = qL$

$$\int D\, dS \cos\theta = q_{enc.}$$

$\therefore$ $D \int dS = q_{enc.}$ $(\because \theta = 0)$

$\therefore$ $D\, 2\pi r\, L = qL$ or $D = \dfrac{q}{2\pi r}$

In vector form, $\vec{D} = \dfrac{q}{2\pi r}\, \hat{r}$

It must be noted that this formula holds within the insulation and outside it.

Problem 2.8 : *A hollow sphere of radius 'a' is centered at the origin and has a uniform surface charge density σ_s. Using Gauss's law, find $\vec{D}$ inside and outside the hollow sphere.*

Solution : Consider a hollow sphere of radius 'a' metre [Refer Fig. 2.20].

Let us find $\vec{D}$ at a point P situated inside the sphere i.e. $r < a$.

To find electric displacement vector $\vec{D}$ at point P, draw a Gaussian surface through point P. The Gaussian surface is a sphere shown by dotted line in Fig. 2.20.

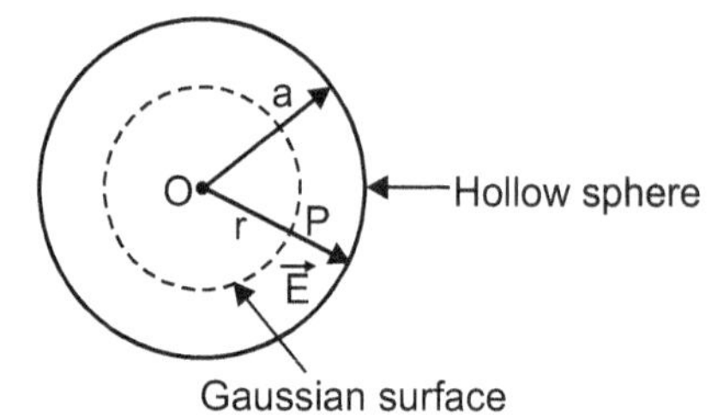

Fig. 2.20

The charge enclosed by Gaussian surface is zero.

$\therefore$ $\int \vec{D} \cdot d\vec{S} = 0$

Thus, we get $D = 0$

So for all points inside the hollow sphere, $D = 0$.

Let us find $\vec{D}$, at a point P situated outside the sphere i.e. $r > a$.

Total charge on the sphere $= \sigma_s\, 4\pi a^2$.

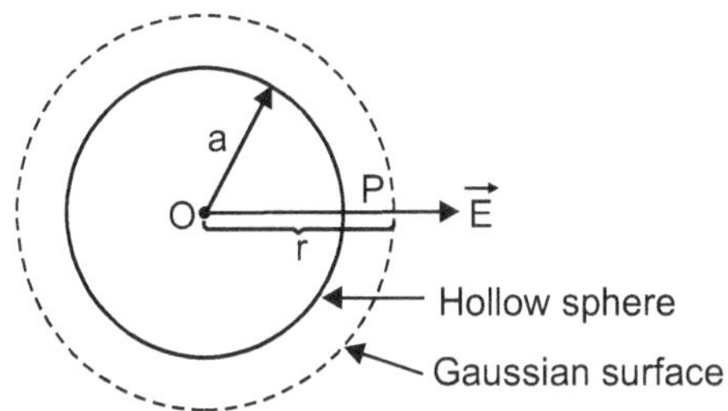

Fig. 2.21

Flux passing through Gaussian surface is,

$$\phi_E = \int \vec{D} \cdot d\vec{S} = \int D \, dS \cos\theta = D \int dS \qquad (\because \theta = 0)$$

$$= D \, 4\pi r^2$$

According to Gauss's theorem, $D \, 4\pi r^2 = q_{enc}$

But, $\qquad\qquad q_{enc} = \sigma_s \, 4\pi a^2$

$$\therefore \qquad D \, 4\pi r^2 = \sigma_s \, 4\pi a^2 \quad \text{or} \quad D = \sigma_s \left(\frac{a}{r}\right)^2$$

In vector form, $\qquad \vec{D} = \sigma_s \left(\frac{a}{r}\right)^2 \hat{r}$

Think Over It

1. What is the effect of dielectric on capacity of parallel-plate capacitor ?

2. Why is there lightning occur in rainy season ?

Summary

- **Electric dipole :** Two equal and opposite charges separated by a small distance is called as an electric dipole.

- **Electric dipole moment :** The product of magnitude of either charge and the distance between the charges is called electric dipole moment. It is denoted by $\vec{P}$ and it is given by

$$\vec{P} = qd$$

SI unit of electric dipole moment is coulomb-meter.

- Potential at a distance r due to an electric dipole is given by

$$V = \frac{1}{4\pi\varepsilon_o} \frac{P \cos\theta}{r^2}$$

- The electric intensity (E) at a distance r due to an electric dipole is

$$E = \frac{P}{4\pi\varepsilon_o r^3} \sqrt{3\cos^2\theta + 1}$$

- Torque on a dipole placed in an electric field is $\vec{\tau} = \vec{P} \times \vec{E}$
  Magnitude of torque is $\tau = PE \sin\theta$ and SI unit of torque is Nm.

- **Non-polar molecule :** A non-polar molecule is a molecule in which effective centre of gravity of the positive charges coincides with effective centre of gravity of negative charges. Problems of non-polar molecule are H_2, N_2, CO_2 etc. This type of molecule does not possess permanent dipole moment.

- **Polar molecule :** It is a molecule in which effective centre of gravity of the positive charges does not coincide with the effective centre of gravity of negative charges. Problems of polar molecule are H_2O, HCl, CO etc. This type of molecule possesses permanent dipole moment.

- **Polarization of dielectric material :** When a dielectric material is placed in an external electric field, a net dipole moment in the direction of field is produced. This phenomenon is known as electric polarization of material.

- **Electric polarization vector $(\vec{P})$:** The net electric dipole moment per unit volume of a dielectric material is known as electric polarization vector or polarization density. It is denoted by $\vec{P}$ and is related with induced charges only.

  The magnitude of $\vec{P}$ is $P = \dfrac{q'}{A}$

  SI unit of electric polarization vector is C/m^2.

- **Gauss's law in dielectrics :** It states that the surface integral of electric displacement vector $\vec{D}$ over a closed surface is equal to free charges enclosed by the closed surface. Mathematically, it can be expressed as $\oint \vec{D} \cdot d\vec{S} = q$

- **Relation between three electric vectors $\vec{D}$, $\vec{E}$ and $\vec{P}$** : The relation between $\vec{D}$, $\vec{E}$ and $\vec{P}$ is

$$\vec{D} = \varepsilon_0 \vec{E} + \vec{P}$$

where (i) $\vec{D}$ is electric displacement vector and is related with free charges only. Its SI unit is C/m^2. Mathematically, magnitude of $\vec{D}$ is $D = \dfrac{q}{A}$

(ii) $\vec{P}$ is polarization vector and it is related with induced charges only. Its S.I. unit is C/m^2. Mathematically, magnitude of $\vec{P}$ is

$$P = \dfrac{q'}{A}.$$

(iii) $\vec{E}$ is electric field vector and it is related with all charges. Its S.I. unit is N/C or volt/meter.

- **The relation between $\vec{P}$ and $\vec{E}$ is** $\vec{P} = \varepsilon_0 (k - 1) \vec{E}$
  where k is dielectric constant and it can be defined as

$$k = \dfrac{\varepsilon}{\varepsilon_0}$$

or $\qquad k = \dfrac{E_0}{E} = \dfrac{V_0}{V} \qquad (\because E_0 = V_0 d \text{ and } E = Vd)$

Exercises

(A) Multiple Choice Questions :

1. The SI unit of electric dipole moment is ......
   (a) C/m^2
   (b) C-m
   (c) C/m
   (d) $C\text{-}m^2$

2. The torque ($\vec{\tau}$) on a dipole placed in an electric field is ......
   (a) $\vec{\tau} = \vec{P} \times \vec{E}$
   (b) $\vec{\tau} = \vec{P} \cdot \vec{E}$
   (c) $\vec{\tau} = PE \sin\theta$
   (d) $\vec{\tau} = 0$

3. The SI unit of electric polarization ($\vec{P}$) is ......
   (a) C-m^2 (b) C/m
   (c) C/m^2 (d) C-m

4. When a dielectric is placed in an external electric field, the field within the dielectric ......
   (a) reduces   (b) increases
   (c) becomes zero   (d) will not change

5. Electric displacement vector ($\vec{D}$) is related with ....
   (a) induced charges only   (b) all charges
   (c) free charges only   (d) none of these

6. The relation between electric vectors $\vec{D}$, $\vec{E}$ and $\vec{P}$ is ......
   (a) $\vec{D} = \vec{P} \cdot \vec{E}$ (b) $\vec{D} = \vec{P} + \vec{E}$
   (c) $\vec{E} = \varepsilon_o \vec{D} + \vec{P}$ (d) $\vec{D} = \varepsilon_o \vec{E} + \vec{P}$

7. Potential at any point on the equatorial line of dipole is ......
   (a) $V = 0$ (b) $V = \infty$
   (c) $V = \dfrac{1}{4\pi\varepsilon_o} \dfrac{q}{r}$ (d) $V = \dfrac{1}{4\pi\varepsilon_o} \dfrac{p}{r^3}$

Answers

1. (b)	2. (a)	3. (c)	4. (a)	5. (c)	6. (d)	7. (a)

(B) State True or False Questions :

1. The non-polar molecule possesses permanent dipole moment.

2. When a dielectric material is placed in an external electric field then a net dipole moment in the direction of field is produced.

3. Electric polarization ($\vec{P}$) is related with free charges only.

4. The SI unit of electric displacement vector ($\vec{D}$) is C/m.

5. The direction of torque is always perpendicular to the plane of dipole axis and electric field.

6. In the presence of dielectric the induced charges (q') due to polarization is always greater than the free charges.

Answers

1. False	2. True	3. False	4. False	5. True	6. False

(C) Short Answer Type Questions :

1. Define the terms electric dipole and electric dipole moment.
2. Define electric dipole moment. Give its SI unit.
3. Define polar molecule. Give examples.
4. Define non-polar molecule. Give examples.
5. Compare polar and non-polar molecules.
6. What do you mean by dielectric material?
7. Define electric polarization vector $\vec{P}$. Give its SI unit.
8. Define electric displacement vector $\vec{D}$. Give its SI unit.
9. Write the relation between three electric vectors $\vec{D}$, $\vec{E}$ and $\vec{P}$.
10. State Gauss's law in dielectrics.
11. What do you mean by polarization of dielectric material ?
12. Define dielectric constant of the material.

(D) Long Answer Type Questions :

1. Explain polar and non-polar molecules with examples and explain the effect of electric field on them.
2. Show that the electric field within the dielectric reduces by a factor k than the external field.
3. Write a note on electric polarization of the dielectric.
4. What is electric dipole and dipole moment ? Obtain an expression for electric potential at any point due to an electric dipole.
5. Derive an expression for electric intensity at any point due to electric dipole.
6. Obtain an expression for torque on a dipole placed in an uniform electric field.
7. State and prove Gauss's law in dielectrics.
8. Define electric vectors $\vec{D}$, $\vec{E}$ and $\vec{P}$. Write the relation between them.
9. Write the relation between three electric vectors $\vec{D}$, $\vec{E}$ and $\vec{P}$. Give the meaning and SI unit of each term.

10. Show that in the presence of dielectric, the induced charge q' due to polarization is always less than the free charge q and is given by $q' = q\left(1 - \dfrac{1}{k}\right)$

(E) Unsolved Problems :

1. The dipole moment of a short electric dipole is 2×10^9 C-m. What would be the electric potential at a distance of 2 metres from the centre of the dipole along the line making an angle of 60° with the axis of the dipole ?   (**Ans.** V = 2.25 volt)

2. An electric dipole consists of two opposite charges each of magnitude 1 μC separated by a distance of 2.0 cm. The dipole is placed in an uniform electric field of intensity 2×10^5 N/C. Calculate the maximum torque on the dipole. (**Ans.** 4×10^{-3} N-m)

3. Calculate the potential and electric intensity due to a dipole of dipole moment 4×10^{-10} C-m at a distance of 2 metre from it on its axis.   (**Ans.** 0.9 V and 0.9 V/m)

4. A parallel-plate capacitor has surface charge density of 10^{-8} C/m^2 on it. The area of each plate is 100 cm^2, and distance between the plates is 1 cm. A glass slab of same shape as that of the capacitor plates and 5 mm thick is kept midway between the plates. Find the three electric vectors in the glass slab.

  (**Ans.** D = 10^{-8} C/m^2, E = 0.113×10^3 V/m, P = 0.9×10^{-8} C/m^2)

5. A parallel-plate capacitor of plate area 100 cm^2 and separation 1.5 cm is charged by a potential of 60 volts. Then the battery is disconnected and a dielectric slab of thickness 0.8 cm and dielectric constant 5 is inserted. Calculate the magnitude of three vectors $\overrightarrow{D}$, $\overrightarrow{E}$ and $\overrightarrow{P}$ in the dielectric.

  (**Ans.** E = 8×10^2 V/m, D = 3.54×10^{-8} C/m^2, P = 2.83×10^{-8} C/m^2)

❑❑❑

Chapter 3...

Magnetization

Contents ...

Pierre Curie (15 May 1859 – 19 April 1906) was a French physicist, a pioneer in crystallography, magnetism, piezoelectricity, and radioactivity. In 1903, he received the Nobel Prize in Physics with his wife, Marie Skłodowska-Curie, and Henri Becquerel, "in recognition of the extraordinary services they have rendered by their joint researches on the radiation phenomena discovered by Professor Henri Becquerel".

Pierre Curie

3.1 Introduction to Magnetization

- In the topic, Dielectrics, we have seen that when a dielectric material or slab is kept in an external electric field, it gets polarized. In the process of polarization, equal and opposite charges are induced on opposite faces of the dielectric slab and due to this, material acquires a resultant electrical dipole moment. The dipole moment per unit volume is called polarization or polarization density ($\vec{P}$).

- Now, we come across a similar situation in magnetism. If a magnetic material is kept in external magnetic field, it gets magnetized and material acquires a resultant magnetic dipole moment. The magnetic dipole moment per unit volume is called magnetization. Let us study, how magnetization occur in material ?

- We know that sources of all magnetic field is electric current, whether it be the current in a wire or current produced by motion of charges within the atoms or molecules of a material. A given material is composed of atoms. Each atom consists of electrons those move in orbits around the nucleus as well as spin about their own axes. The circulating electrons are equivalent to current loops. Such circulating currents were first postulated by Ampere and is called Amperian currents or atomic currents.

The current carrying element of very small dimension (Fig. 3.1) can be considered as magnetic dipole. So in a given material, each current loop can be considered as a magnetic dipole and hence it has magnetic dipole moment or magnetic moment.

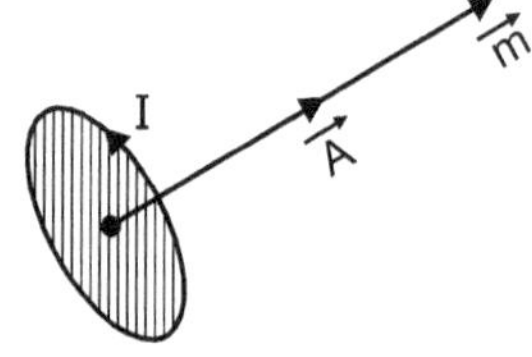

Fig. 3.1

The magnetic dipole moment is given by

$$\vec{m} = I\vec{A} \qquad \qquad \text{... (3.1)}$$

where, I is the current flowing through a loop of area A.

S.I. unit of magnetic dipole moment is ampere - (meter)^2. Due to spin and orbital motion of electron, each electron has spin magnetic moment and orbital magnetic moment. The vector sum of all such magnetic moments of electrons gives resultant magnetic moment to each atom.

- In the absence of external magnetic field, magnetic moments of atoms are randomly oriented, so there is no net magnetic moment in any volume of material, hence material does not possess any magnetic field.

- When the material is kept in an external magnetic field, magnetic moments of atoms more or less align themselves with the direction of external magnetic field. So the net magnetic dipole moment of material is not zero and the material is said to be magnetized.

- The effect of alignment of atomic dipole moment in external magnetic field is described by a quantity called magnetization ($\vec{M}$). *It is defined as a magnetic dipole moment per unit volume.* If there are N atoms in a given volume ΔV and i^{th} atom has magnetic moment m_i then magnetization is defined as

$$\vec{M} = \lim_{\Delta V \to 0} \frac{\sum_{i=1}^{N} \vec{m_i}}{\Delta V} \qquad \ldots (3.2)$$

The S.I. unit of magnetization M can be given as

$$\text{Magnetization} = \frac{\text{Magnetic dipole moment}}{\text{Volume}}$$

$$= \frac{\text{Current} \times \text{Area}}{\text{Volume}} = \frac{\text{A-m}^2}{\text{m}^2}$$

Thus, S.I. unit of magnetization is ampere/meter (A/m).

3.2 Magnetic Materials

- Magnetic materials and magnets are used in many applications such as refrigerator, VCR, audio cassettes, speakers and microphones, MRI in hospitals, magnetic storage devices like hard disks. The coils in transformers, motors, generators and electromagnets always have iron cores to increase the magnetic field.
- Depending upon the behaviour of material under external magnetic field, materials are classified under following categories :
  (i)   Diamagnetic materials.
  (ii) Paramagnetic materials.
  (iii) Ferromagnetic materials.
  (iv) Anti-ferromagnetic materials.
  (v)   Ferri-magnetic materials.

3.3 Types of Magnetic Materials

3.3.1 Diamagnetic Materials

- The material which is weakly repelled by magnetic field is called diamagnetic material.
- In some materials, the individual atoms do not have a net magnetic dipole moment. When such material is placed in an external magnetic field, the external field alters electron motions within the atoms, causing additional current loops and dipole moments are induced in

the atoms by the applied field. The resultant field in such material is, therefore, less than the applied field. This phenomenon is called *diamagnetism,* and these materials are called *diamagnetic* materials. Thus, when diamagnetic materials are placed in external magnetic field, they are weakly repelled.

Examples of such substances are bismuth, antimony, gold, water, alcohol, quartz, hydrogen, mercury, copper etc.

Properties of Diamagnetic Substances :

1. The individual atoms of diamagnetic substances do not have permanent dipole moment.

2. A diamagnetic substance is weakly repelled by a magnetic field.

3. If a thin rod of diamagnetic material is placed in a uniform magnetic field, it comes to rest with its length perpendicular to the direction of magnetic field.

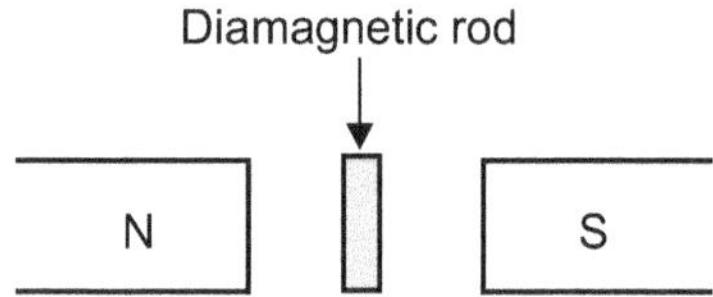

Fig. 3.2 : Diamagnetic rod in magnetic field

4. Magnetic field lowers the level of diamagnetic liquid when subjected to magnetic field to its arm in 'U' shaped tube.

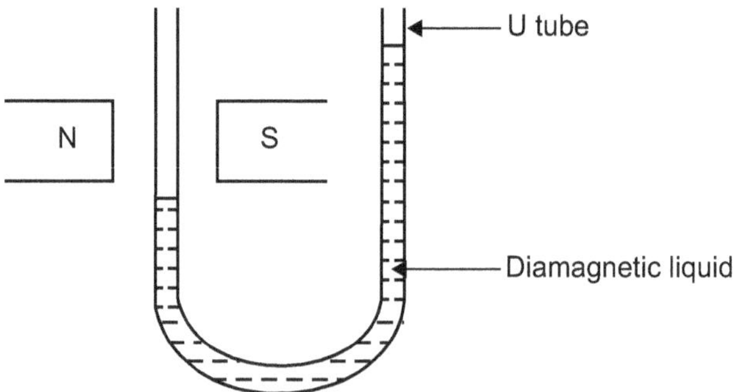

Fig. 3.3 : Diamagnetic liquid in U tube in magnetic field

5. When placed in a non-uniform magnetic field, a diamagnetic substance has tendency to move from the stronger part to weaker part of the field.

6. When such material is placed in an external field, it is magnetized opposite to the field direction.

3.3.2 Paramagnetic Materials

- The material which is weakly attracted by magnetic field is called paramagnetic material.

- In paramagnetic materials, each atom possesses net dipole moment. In materials, the dipole moments of atoms are randomly oriented giving out net dipole moment equal to zero. When such material is placed in an external magnetic field, the atomic dipole moments are aligned along the direction of the applied field. The resultant magnetic field in the material is greater than the applied field. The tendency of the material to increase the magnetic field due to magnetization is called *paramagnetism* and the materials are called paramagnetic materials.

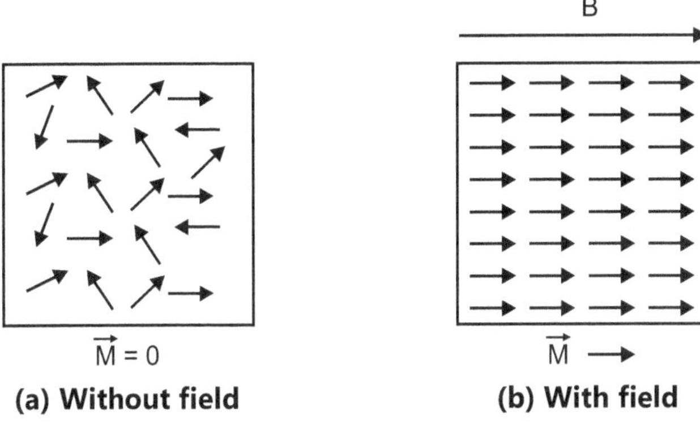

(a) Without field   **(b) With field**

Fig. 3.4 : Paramagnetic substance

Examples of such materials are aluminium, platinum, manganese, oxygen etc.

Properties of Paramagnetic Substances :

1. A paramagnetic substance is weakly attracted by a magnetic field.

2. If a thin rod of paramagnetic material is placed in a uniform magnetic field, it comes to rest with its length along the direction of magnetic field.

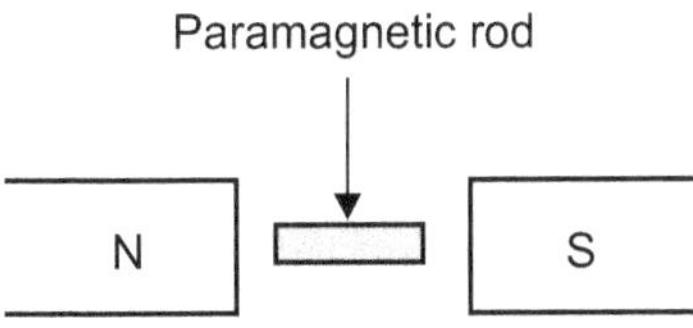

Fig. 3.5 : Thin paramagnetic rod in external magnetic field

3. Magnetic field increases the level of paramagnetic liquid when subjected to magnetic field to its arm in 'U' shaped tube.

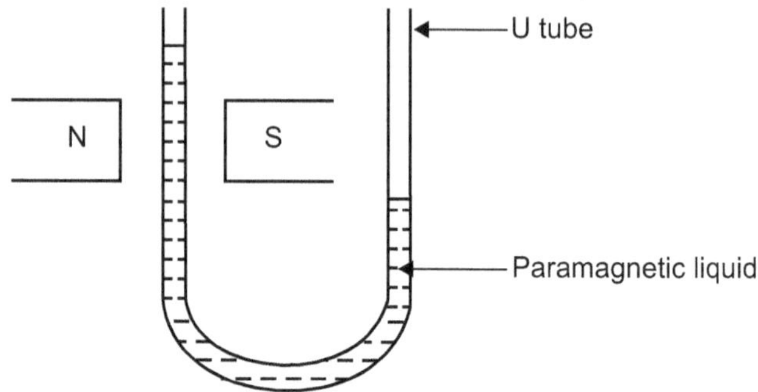

Fig. 3.6 : Paramagnetic liquid in U tube in magnetic field

4. When placed in a non-uniform magnetic field, a paramagnetic substance has tendency to move from the weaker part to stronger part of the field.

5. When such material is placed in an external magnetic field, it is magnetized along the direction of field.

3.3.3 Ferromagnetic Materials

- The material is strongly attracted by magnetic field is called ferromagnetic material.

- In ferromagnetic materials, each atom has net dipole moment like paramagnetic substances. In these materials, the strong interaction between magnetic dipole moments of atoms cause them to line up parallel to each other in small regions called *domain*, even if there is no external magnetic field. The explanation for this co-operative effect requires quantum mechanics which is beyond the scope of the book. Each domain has a net magnetization. Typical domain size is about 1 mm containing about 10^{11} atoms in each domain.

- Fig. 3.7 (a) shows an example of magnetic domain structure. Within each domain, nearly all the magnetic moments of atoms are parallel. When there is no external magnetic field, the domain magnetizations are randomly oriented and there is no net magnetization (M = 0).

- When external magnetic field is applied, the domains tend to orient themselves parallel to the field. In external magnetic field, the domains which are aligned along the direction of the field grow in size and those are magnetized in other directions get reduced in size and domain boundaries are shifted [Refer Fig. 3.7 (b)].

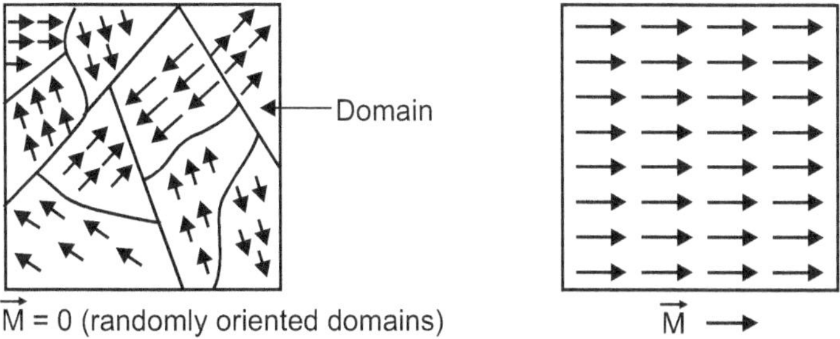

(a) Without external magnetic field   **(b) With external magnetic field**

Fig. 3.7

- As external field is increased, a point is reached where all magnetic moments in the ferromagnetic material are aligned parallel to the external field. This condition is called **saturation magnetization.** After this stage, there is no increase in magnetization. Fig. 3.8 shows a magnetization curve (M versus magnetic field B).

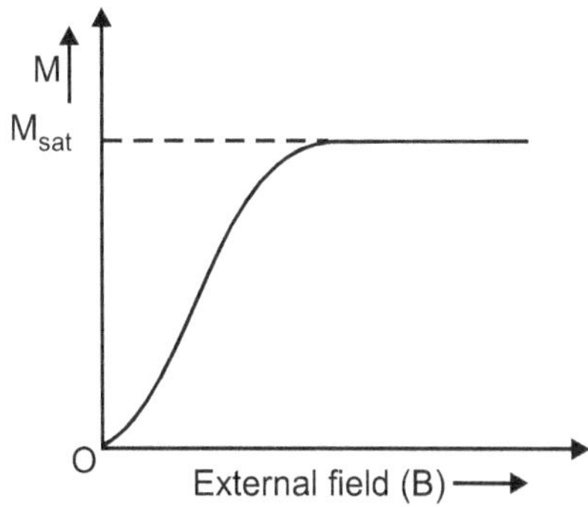

Fig. 3.8 : Magnetization curve for a ferromagnetic material

- In some ferromagnetic materials, the magnetization persists even after external field is removed. Such materials are called hard magnetic materials or hard ferromagnets. Alnico (an alloy of iron, aluminium, nickel, cobalt) is one such material. The naturally occurring lodestone is another. Such materials are used to make permanent magnets.

- On the other hand, there is a class of ferromagnetic materials in which the magnetization disappears on the removal of external field. Such materials are called soft ferromagnetic materials. Examples of soft ferromagnetic materials are soft iron, nickel, gadolinium.

- The ferromagnetic property depends on temperature.

- At high enough temperature, ferromagnetic substances become paramagnetic. The domain structure disintegrates with temperature. The disappearance of magnetization with temperature is gradual. The temperature of transition from ferromagnetism to paramagnetism is called the **Curi temperature (T_C)**.

- Following Table 3.1 shows Curi temperature (T_C) of some ferromagnetic materials.

Table 3.1

Material	T_C (K)
Cobalt	1394
Iron	1043
Fe_2O_3	893
Nickel	631
Gadolinium	317

3.3.4 Antiferromagnetic Materials

- Antiferromagnetic materials show interesting modification of ferromagnetic behaviour. In these materials, the magnetic moments of atoms or molecules, usually related to the spins of electrons, align in a regular pattern with neighbouring spins pointing in opposite directions as shown in Fig. 3.9.

- For example, in antiferromagnetic material such as manganese oxide (MnO), adjacent ions behave as a tiny magnet, spontaneously align themselves at relatively low temperature in anti-parallel manner throughout the material (Refer Fig. 3.9). The magnetic moments from ions or atoms oriented in one direction are cancelled out by the set of magnetic moments of ions or atoms that are aligned in the reverse direction. So the material exhibits no net magnetism.

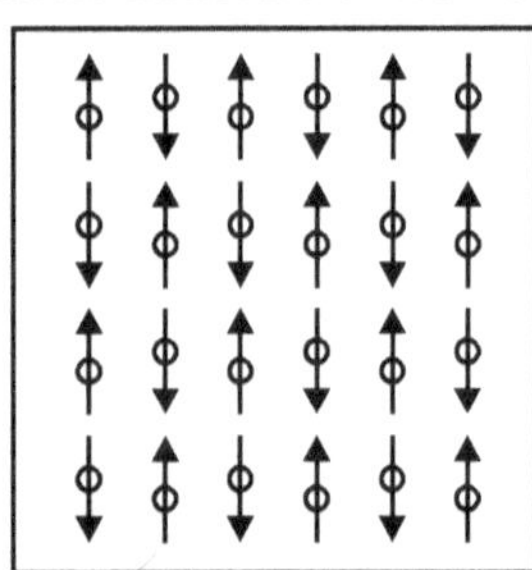

Fig. 3.9 : Antiferromagnetism

- This antiparallel coupling of magnetic moments disappear entirely above a certain temperature called Neel temperature which is the characteristic of antiferromagnetic material. For example, the Neel temperature for MnO is 122 K.
- When an external magnetic field is applied, the material behaves similar to paramagnetic material.
- In antiferromagnetic material, χ_m increases with temperature while in paramagnetic material, χ_m decreases with temperature. Antiferromagnetic materials are commonly among transition metal compounds, especially oxides. The difference between paramagnetic and diamagnetic substances, and ferromagnetic and paramagnetic substances are given in tabular form.

- **Difference between Paramagnetic and Diamagnetic Substances :**

Paramagnetic Substance	Diamagnetic Substance
1. It is weakly attracted by magnetic field.	1. It is weakly attracted by magnetic field.
2. Each atom in this material has net magnetic moment.	2. Individual atoms in these materials do not possess magnetic moments.
3. The permeability of this material is greater than the permeability of free space.	3. The permeability of this material is less than the permeability of free space.
4. When such material is placed in the external field, it is magnetized along the direction of field.	4. When such material is placed in the external field, it is magnetized opposite to the direction of field.
5. When paramagnetic substance is suspended in non-uniform magnetic field, it has tendency to move from weaker to stronger part of the field. e.g. Aluminium, Platinum, Manganese, Oxygen.	5. When diamagnetic substance is suspended in non-uniform magnetic field, it has tendency to move from stronger to weaker part of the field. e.g. Bismuth, Copper, Gold, Mercury.

- **Difference between Ferromagnetic and Diamagnetic Substances :**

Ferromagnetic Substance	Diamagnetic Substance
1. It is strongly attracted by magnetic field.	1. It is weakly attracted by magnetic field.
2. Each atom in this material has net magnetic moment.	2. Individual atoms in these materials do not possess magnetic moments.

... Contd.

3. The substance is made up of small domains. All magnetic moments of atoms in one domain are aligned along the same direction.	3. Domain structure does not exist.
4. The permeability of this material is very large as compared to the permeability of free space.	4. The permeability of this material is less than the permeability of free space.
5. When such materials are placed in the external field, they get strongly magnetized along the direction of field even in weak field. e.g. Iron, Nickel and Cobalt.	5. When such material is placed in the external field, it is magnetized opposite to the direction of field. e.g. Bismuth, Copper, Gold, Mercury.

3.4 Bohr Magneton

- An electron magnetic dipole moment can be expressed in terms of physical constant. A physical constant and the natural unit of expressing an electron magnetic dipole moment is called Bohr magneton (μ_B).

- Let us obtain expression for Bohr magneton (μ_B) using loop model for electron orbit.

- We know that atoms contain electrons and these electrons form microscopic current loops that produce magnetic field. In many materials, these currents are randomly oriented and they do not cause any net magnetic field. If field is produced outside some material, the current loops are oriented in the direction of fields. The orientation causes increase in their magnetic fields. Then we say that the material is magnetized.

- Consider an electron of mass m and charge e moving with speed $\vec{v}$ in a circular orbit of radius r as shown in Fig. 3.10.

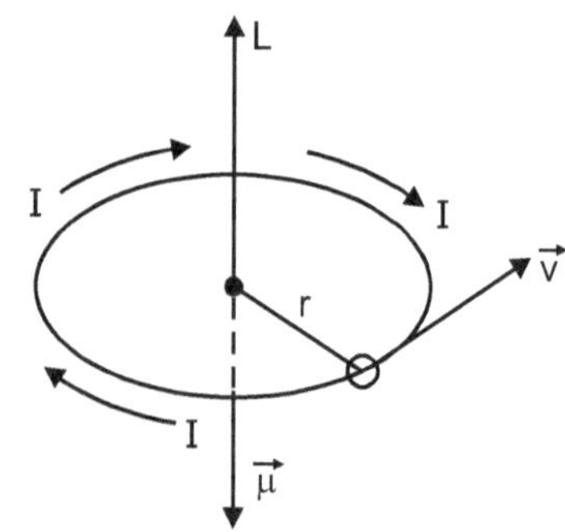

Fig. 3.10 : Electron in an atom

This moving charge is equivalent to a current loop. The magnetic dipole moment or magnetic moment (μ) of the loop is given by

$$\mu = IA \qquad \qquad \text{... (3.4)}$$

where I – current in loop and A – area of loop (πr^2)

(The direction of $\vec{\mu}$ is given by right hand rule. Wrap the fingers of your right hand around the perimeter of loop in the direction of current. Extend your thumb so that it is perpendicular to the plane of the loop.

The direction of thumb indicates the direction of $\vec{\mu}$.)

The period of orbit T, is given by,

$$T = \frac{2\pi}{\omega} = \frac{2\pi r}{v} \qquad (\text{as } v = r\omega)$$

The current I is the total charge passing per unit time.

$$I = \frac{e}{T} = \frac{ev}{2\pi r} \qquad \qquad \text{... (3.5)}$$

The magnetic moment μ is given by,

$$\mu = IA = \frac{ev}{2\pi r} \cdot (\pi r^2) = \frac{evr}{2} \qquad \qquad \text{... (3.6)}$$

It is useful to express μ in terms of the angular momentum L of electron.

The angular momentum for a particle moving in circular path is the product of magnitude of momentum (mv) and radius of orbit.

i.e. $L = mvr$ ... (3.7)

Using equation (3.7) in equation (3.6), we express μ in terms of L.

$$\mu = \frac{ev}{2} \times \frac{L}{mv}$$

$$\mu = \left(\frac{e}{2m}\right) L \qquad \qquad \text{... (3.8)}$$

Equation (3.8) is useful because atomic angular momentum is quantized. The component of angular momentum in a particular direction is always integer multiple of $h/2\pi$, where h is a constant called as Planck's constant. The magnitude of h is

$$h = 6.626 \times 10^{-34} \text{ Js} \sim 6.630 \times 10^{-34} \text{ J·s}$$

The quantity $\dfrac{h}{2\pi}$ represents a fundamental unit of angular momentum in atomic system just as e is a fundamental unit of charge.

Equation (3.8) shows that associated with the fundamental unit of L there is a corresponding fundamental unit of μ.

$$\text{If } L = \frac{h}{2\pi} \text{ then } \mu = \frac{e}{2m}\left(\frac{h}{2\pi}\right)$$

$$\therefore \qquad \mu_B = \frac{eh}{4\pi m}$$

This quantity μ_B is called as **Bohr magneton**. Thus by classical (non-quantum) analysis, we have obtained same result, as given by quantum physics.

Substituting the values of e, h and m, we obtain numerical value of Bohr magneton.

$$\mu_B = \frac{(1.6 \times 10^{-19}) \times 6.630 \times 10^{-34}}{4 \times 3.14 \times 9.1 \times 10^{-31}}$$

$$= 9.274 \times 10^{-24} \text{ A.m}^2 \approx \mathbf{9.27 \times 10^{-24} \text{ J/T}}$$

Similarly, spin magnetic dipole moments of electrons and other elementary particles can be expressed in terms of Bohr magneton μ_B.

Definition-Bohr magneton : It is a physical constant and the natural unit for expressing an electron magnetic dipole moment. The Bohr magneton is defined in S.I. unit by

$$\mu_B = \frac{e}{2m}\left(\frac{h}{2\pi}\right) \quad \text{or} \quad \mu_B = \frac{e\hbar}{2m_e} \qquad \qquad \text{... (3.9)}$$

where e is the charge of electron, $\hbar$ is the reduced Planck constant $(h/2\pi)$ and m_e is the electron rest mass

For an electron, the magnitude of measured z component of $\vec{\mu_s}$ is

$$\mu_{s,z} = 1\,\mu_B$$

The magnitude of an electron's spin magnetic moment is approximately one Bohr magneton.

Solved Problem

Problem 3.1 : *A domain in ferromagnetic iron is in the form of a cube of side length 1 mm. Estimate the number of iron atoms in the domain and the maximum possible dipole moment and magnetization of the domain. The molecular mass of iron is 55 g/mole and its density is 7.9 g/cm³. Assume that each iron atom has a dipole moment of 9.27×10^{-24} A m².*

Solution : Given : $l = 1$ mm $= 1 \times 10^{-6}$ m, $\rho = 7.9$ g/cm³,

$$\mu = 9.27 \times 10^{-24} \text{ A m}^2.$$

The volume of the cubic domain is

$$v = (1 \times 10^{-6})^3 = 1 \times 10^{-18} \text{ m}^3 = 10^{-12} \text{ cm}^3$$

Its mass is volume × density.

$$m = 7.9 \times 10^{-12} \text{ g}$$

It is given that an Avogadro number (6.023×10^{23}) of iron atoms has a mass of 55 g. Hence, the number of atoms in the domain is

$$N = \frac{7.9 \times 10^{-12} \times 6.023 \times 10^{23}}{55} = \mathbf{8.65 \times 10^{10} \text{ atoms}} \quad \textbf{... Ans.}$$

The maximum possible dipole moment μ_{max} is achieved for the (unrealistic) case when all the atomic moments are perfectly aligned. Thus,

$$\mu_{max} = N\mu = (8.65 \times 10^{10}) \times (9.27 \times 10^{-24})$$
$$= 8.02 \times 10^{-13} \text{ Am}^2 \approx \mathbf{8 \times 10^{-13} \text{ Am}^2} \qquad \textbf{... Ans.}$$

The consequent magnetization is

$$M_{max} = \frac{\mu_{max}}{\text{Domain volume}} = \frac{8 \times 10^{-13}}{1 \times 10^{-18}} = \mathbf{8 \times 10^{5} \text{ Am}^{-1}} \qquad \textbf{... Ans.}$$

Think Over It

1. What is the origin of magnetic moment in magnetic materials ?
2. What is the main difference between ferromagnetic and antiferromagnetic materials ?

Summary

- Any current carrying element of very small dimension can be considered as a magnetic dipole.

  The magnetic dipole moment is given by

  $$\vec{m} = I\vec{A}$$

  where I is the current flowing through a loop of area A.

  The S.I. unit of magnetic dipole moment is ampere (meter)^2.

- **Magnetization ($\vec{M}$) :** It is defined as the magnetic dipole moment per unit volume. The S.I. unit of magnetization is ampere/meter.

- **Diamagnetism :** When a material is placed in an external magnetic field, the external field alters electron motions within the atoms, causing additional current loop and dipole moments are induced in the atoms by the applied field. The resultant field in such materials is therefore less than applied field. This phenomenon is called diamagnetism and these materials are called diamagnetic materials. Examples of such materials are bismuth, antimony, gold, water, alcohol, copper etc.

- **Paramagnetism :** When a material is placed in an external magnetic field, the atomic dipole moments are aligned along the direction of the applied field. The resultant magnetic field in the material is greater than the applied field. The tendency of the material to increase the magnetic field due to magnetization is called paramagnetism and the materials are called paramagnetic materials.

  Examples of such materials are aluminium, platinum, manganese, etc.

- **Ferromagnetic materials :** The material is made up of small domains. A domain is a small region where magnetic moments of atoms are aligned along the same direction. In the absence of external field, magnetic moments of domains are randomly oriented giving out net magnetic moment zero. When external magnetic field is applied, the magnetic moments of domains tend to orient themselves parallel to the field and the material acquires net magnetization along the direction of field. Such materials are called ferromagnetic materials. Examples of ferromagnetic materials are nickel, cobalt, iron etc.

- **Antiferromagnetic materials :** In these materials, adjacent ions or atoms have identical magnetic moments but oppositely directed, so magnetic moments in one direction is cancelled out by the set of magnetic moments of ions or atoms that are aligned in the reverse direction. Hence the material exhibits no net magnetism. When such material is kept in external magnetic field, the material behaves similar to paramagnetic material. Example is MnO.

- **Bohr magneton :** It is a physical constant and the natural unit for expressing an electron magnetic dipole moment. The Bohr magneton in S.I. unit is given by $\mu_B = \dfrac{e\hbar}{2m_e}$ where m_e is mass of an electron and $\hbar = \dfrac{h}{2\pi}$.

Exercises

(A) Multiple Choice Questions :

1. The S.I. unit of magnetic dipole moment is ......
   - (a) ampere-meter
   - (b) ampere-(meter)^2
   - (c) ampere/meter
   - (d) ampere/(meter)^2

2. The Bohr magneton is defined in S.I. unit by ......
   - (a) $\dfrac{eh}{2m_e}$
   - (b) $\dfrac{e\hbar}{2m_e}$
   - (c) $\dfrac{eh}{2\pi m}$
   - (d) $\dfrac{e\hbar}{2m_e c}$

3. A domain structure exists in ......
   - (a) diamagnetic substances
   - (b) paramagnetic substances
   - (c) ferromagnetic substances
   - (d) antiferromagnetic substances

4. When a diamagnetic substance is kept in external magnetic field, it gets magnetized ......
   - (a) weakly along the direction of external field
   - (b) strongly along the direction of external field
   - (c) opposite to the direction of external field
   - (d) none of these

5. In ...... material adjacent ions or atoms have identical but oppositely directed magnetic moments.
   - (a) paramagnetic
   - (b) diamagnetic
   - (c) ferromagnetic
   - (d) antiferromagnetic

Answers

1. (b)	2. (b)	3. (c)	4. (c)	5. (d)

(B) State True or False Questions :

1. The S.I. unit of magnetization is ampere/meter.
2. Aluminium is a diamagnetic substance.
3. A paramagnetic substance is weakly attracted by a magnetic field.
4. A diamagnetic substance is strongly repelled by a magnetic field.
5. In hard ferromagnetic material, magnetization disappears on the removal of external field.

Answers

1. True	2. False	3. True	4. False	5. False

(C) Short Answer Type Questions :

1. Define the term magnetic dipole moment and give its S.I. unit.

2. Define magnetization and give its S.I. unit.

3. Give two sources of magnetic moments for electrons.

4. What do you mean by diamagnetism ?

5. What do you mean by paramagnetism ?

6. What do you mean by domain in ferromagnetic materials ?

7. What do you mean by soft ferromagnetic materials ?

8. What is Neel temperature ?

9. What is Curi temperature ?

10. Give any two examples of paramagnetic substances.

11. Give any two examples of ferromagnetic substances.

12. What is meant by saturation magnetization ?

(D) Long Answer Type Questions :

1. Write a note on magnetization of a magnetic material.

2. Define Bohr magneton and obtain an expression for it.

3. What is a diamagnetic material ? Give properties of diamagnetic materials.

4. What do you mean by paramagnetic materials ? Give properties of paramagnetic materials.

5. Write a note on antiferromagnetic materials.

6. Write a note on ferromagnetic materials.

7. Distinguish between paramagnetic and diamagnetic materials.

8. Distinguish between ferromagnetic and paramagnetic materials.

9. Distinguish between ferromagnetic and diamagnetic materials.

10. Cite the differences between ferro and antiferromagnetic materials.

Chapter 4...

Magnetostatics

Contents ...

Danish physicist and philosopher H. C. Oersted, who, in 1819, discovered the deflection of a compass needle while performing a demonstration for his student. This discovery of a fundamental connection between electricity and magnetism rocked the scientific community. Oersted concluded that moving charges or currents produce a magnetic field in the surrounding space. The CGS unit Oersted was established by the International Electro technical Commission in 1930 in the honour of the Hans Christian Oersted.

Hans Christian Oersted (1777-1851)

Introduction

- We have already discussed electrostatics, which deals with physics of the electric field created by static charges. Electric currents are produced by moving charges undergoing translational motion. Magnetostatics is a branch of electromagnetics involving magnetic fields produced by steady non-time varying currents. The link between electric and magnetic field was first established by Oersted in 1820. He discovered that an electric current flowing through a wire or conductor produces magnetic effects. The main source of all magnetic fields is an electric current. A current or moving charge responds to the magnetic field and so experiences a magnetic force. The magnetic forces play important role in electronic devices like electric motors, T.V. picture tubes, microwave ovens, load-speakers, computer printers and disk drives. A well known MRI (Magnetic Resonance Imaging) technique make it possible to see details of soft tissue that are not visible by X-ray image. Magnetography techniques like MCG (Magnetocardiography) and MEG (Magneto encephallography) are the techniques used in the diagnosis of heart and brain diseases respectively.

- In this chapter we shall mainly consider the magnetic field produced by a direct current. In this we will study some important concept of magnetic field, applications of Biot-Savart's law, Ampere's circuital law and Gauss's law for magnetism.

4.1 Introduction to Magnetization

- We have seen that a charged object produces an electric field at all points in space. In a similar manner, a bar magnet is a source of a magnetic field. The space around the bar magnet is called magnetic field in which force will act on other bar magnet when placed in the field. Unlike electric charges which can be isolated, the two magnetic poles always come in a pair. The poles are called North pole and South pole. There are two ways to create magnetic fields :

1. Moving electric charged particle creates magnetic field.

2. Elementary particles such as protons, neutrons and electrons have intrinsic magnets called magnetic moments also create magnetic field.

The magnetic field is expressed in terms of magnetic field lines. Magnetic field lines were introduced by Michael Faraday (1791-1867) who named them "lines of force." Like electric field, we can draw these lines so that the line at any point is tangent to magnetic field vector $\vec{B}$ as shown in Fig. 4.1.

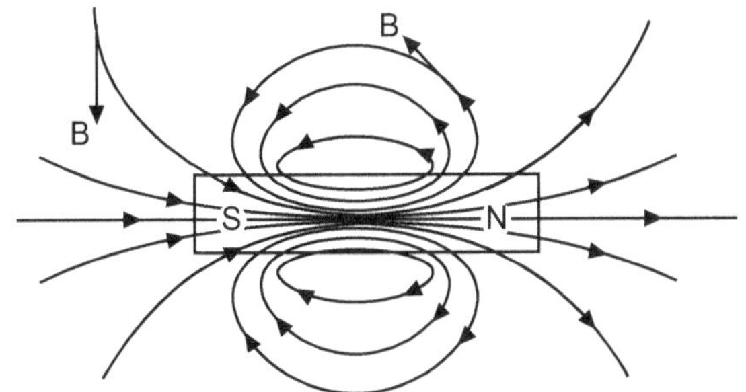

Fig. 4.1 : Magnetic field lines of a permanent magnet

From Fig. 4.1,

1. At each point, the magnetic field vector $\vec{B}$ is tangent to the field line.

2. The spacing of the lines represents the magnitude of $\vec{B}$.

    More densely field lines are packed, indicates stronger field at that point. If field lines are far apart, then field magnitude is small.

3. At each point, the field lines indicate the same direction as magnetic compass would indicate.

4. Magnetic field lines do not intersect. They point away from the north pole and point towards the south pole.

5. The magnetic field lines are not "lines of force", because unlike electric field lines they do not point in the direction of the force on a charge.

- Fig. 4.2 shows the magnetic field lines produced by several sources of magnetic field. From Fig. 4.2 (a) between flat and parallel magnetic poles of C shaped magnet, the magnetic field is nearly uniform. The field lines are approximately straight, parallel and equally spaced showing the magnetic field in this region is uniform.

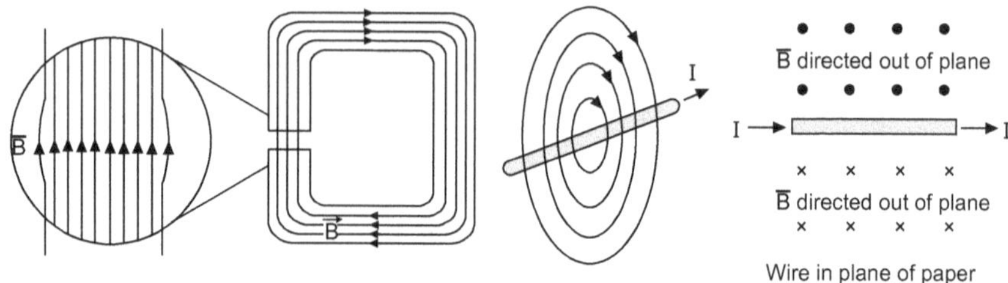

(a) Magnetic field of C shaped magnet

(b) Magnetic field of straight conductor carrying wire

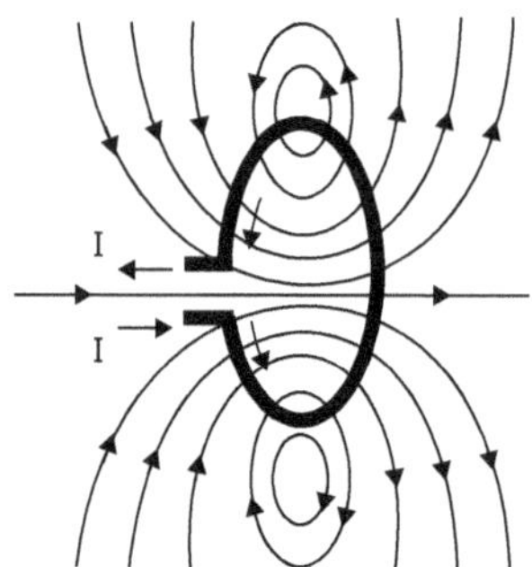

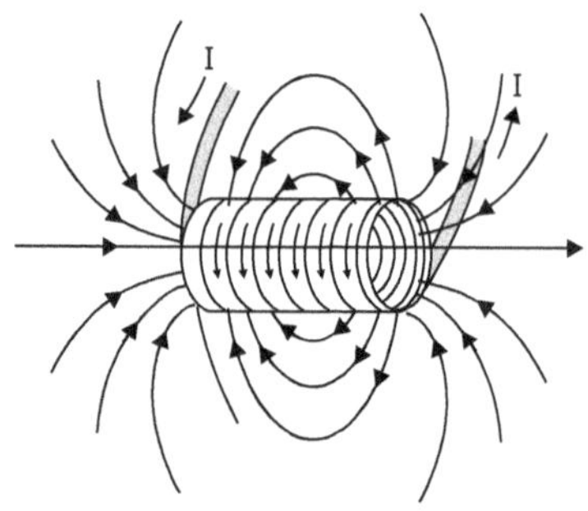

(c) Magnetic field of current carrying loop

(d) Magnetic field of a solenoid

Fig. 4.2 : Magnetic field lines produced by several sources

- Fig. 4.2 (b) shows the magnetic field lines of a straight current carrying wire. Note that to represent a field coming out of or going into the paper, we use dots and crosses respectively. Fig. 4.2 (c) represents magnetic field lines of current carrying loop and Fig. 4.2 (d) a current carrying coil (solenoid). Note that field of the loop and that of coil are like a field of bar magnet as shown in Fig. 4.2. Field lines can be used as a qualitative tool to visualize magnetic forces. In ferromagnetic substances like iron and in plasmas, magnetic forces can be understood by imagining that the field lines exert a tension (like a rubber band) along their length, and a pressure perpendicular to their length on neighbouring field lines. 'Unlike' poles of magnets attract because they are linked by many field lines; 'like' poles repel because their field lines do not meet, but run parallel, pushing on each other.

4.2 Magnetic Induction and Intensity of Magnetization

- The magnetic field can be defined in several equivalent ways based on the effects it has on its environment. The magnetic field B is defined from the Lorentz Force Law, and specifically from the magnetic force on a moving charge.

- *If a positive test charge q is moving with velocity $\vec{v}$ through a point P situated in magnetic field and experiences a deflection force $\vec{F}$, then the magnetic field is defined by the relation*

$$\vec{F} = q\,\vec{v} \times \vec{B}$$

- By measuring the magnetic force $\vec{F}$ acting on q moving with velocity v, we can obtain $\vec{B}$.

- The right hand rule tells us that the thumb of right hand points in the direction of $\vec{v} \times \vec{B}$ when the fingers sweep $\vec{v}$ into $\vec{B}$. If q is positive, then direction of $\vec{F} = q\vec{v} \times \vec{B}$ is in the direction of $\vec{v} \times \vec{B}$. If q is negative, then the direction of $\vec{F}$ is opposite to $\vec{v} \times \vec{B}$.

- The direction of $\vec{F}$ is always perpendicular to the plane containing $\vec{v}$ and $\vec{B}$ as shown in Fig. 4.3.

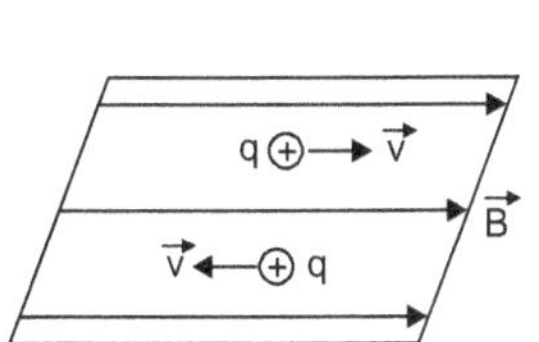

(a) A charge moving parallel to $\vec{B}$, experiences zero magnetic force

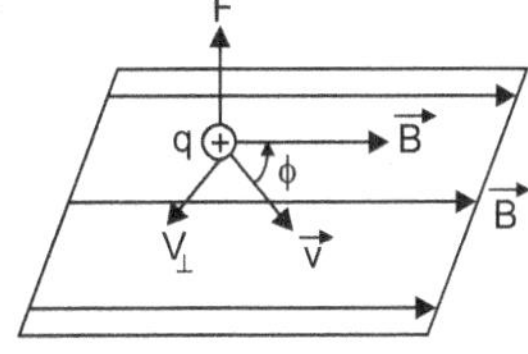

(b) $\vec{F}$ is perpendicular to $\vec{v}$ and $\vec{B}$

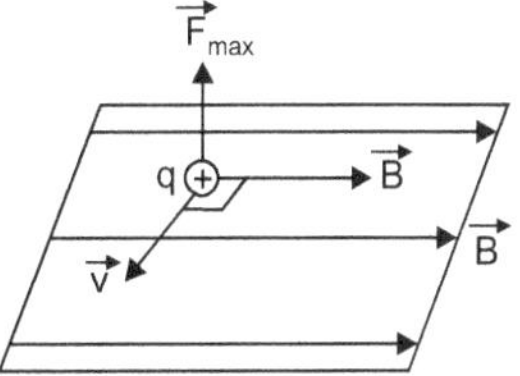

(c) Maximum force of magnitude $F_{max} = qvB$

Fig. 4.3

The magnitude of force is

$$F = q\,v\,B\,\sin\phi \qquad \qquad \text{... (4.1)}$$

$$\text{or} \qquad B = \frac{F}{q\,v\,\sin\phi} \qquad \qquad \text{... (4.2)}$$

where ϕ is the angle between direction of $\vec{v}$ and $\vec{B}$.

If a charge is moving perpendicular to a magnetic field, it will experience a maximum magnetic force with magnitude [Fig. 4.3 (c)]

$$F_{max} = q\,v\,B \qquad \qquad \text{... (4.3)}$$

Thus, a force on a charge q, moving with velocity $\vec{v}$ in the direction of magnetic field B is $\vec{F} = q\vec{v} \times \vec{B}$ ... (4.4)

The magnetic field at a particular point can be described in terms of *magnetic induction* $\vec{B}$.

Unit of $\vec{B}$

If force F is in newton, q in coulomb and v in meter per second, then the S.I. unit of B from equation (4.3) is

$$\frac{\text{newton}}{\text{coulomb meter/second}} \quad \text{i.e.} \quad \frac{N}{C}\frac{1}{m/s}.$$

This unit is given the name weber/(meter)^2 or Wb/m^2. It is also called tesla (T).

$$\therefore \qquad 1 \text{ tesla} = 1\,T = 1\frac{Wb}{m^2} = \frac{N}{C} \cdot \frac{1}{m/s} = 1\frac{N}{A\text{ - }m}$$

A smaller unit (on SI) called gauss (denoted by G) is also often used.

$$\therefore \quad 1\,G = 10^{-4}\,T \text{ or } 1\,T = 10^4 \text{ gauss}$$

The magnetic fields generated by currents or current carrying conductors are characterized by the magnetic field $\vec{B}$ measured in Tesla. But when the generated fields pass through magnetic materials which themselves contribute internal magnetic fields, *i.e.* the field comes from the external currents and what comes from the material itself. It has been common practice to define another magnetic field quantity, usually called the "magnetic field strength or intensity" designated by $\vec{H}$. The relation between $\vec{B}$ and $\vec{H}$ is given as

$$\vec{B} = \mu\vec{H}$$

where μ is called permeability of the material. In free space, we have

$$\overrightarrow{B} = \mu_o \overrightarrow{H}$$

where μ_o is called permeability of free space.

The unit of magnitude of H is A/m.

4.3 Biot-Savart's Law

4.3.1 Statement of Biot-Savart's law

- **Biot-Savart's law** is an equation describing the magnetic field generated by an electric current. It relates the magnetic field to the magnitude, direction, length, and proximity of the electric current. The law is valid in magnetostatics, and is consistent with both Amperes circuital law and Gauss law for electromagnetism. It is named for Jean-Baptiste Biot and Felix Savart who discovered this relationship in 1820.

- A current carrying conductor (or wire) produces a magnetic field around it. The magnetic field, at any point, can be obtained with the help of Biot-Savart's law.

- Consider a small element of length 'dl' of a conductor carrying current 'I' (Refer Fig. 4.4).

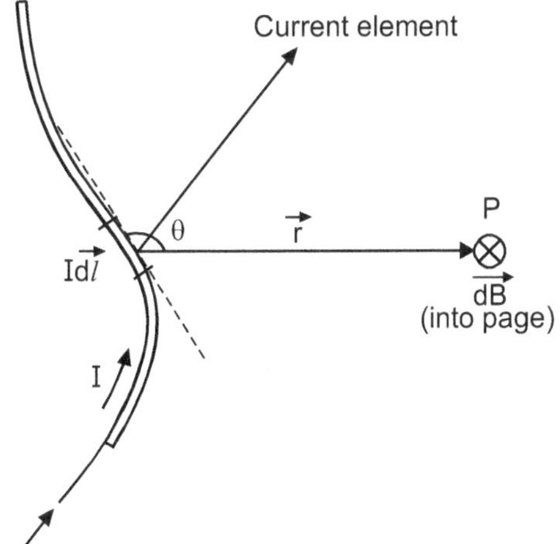

Fig. 4.4 : Illustration of Biot-Savart's law

(A current element Idl produces a magnetic field $\overrightarrow{dB}$ at $\overrightarrow{r}$)

Consider a point P at a distance r from the centre of this element. Let θ be the angle between the direction of current element[*] and the line joining the centre of the element to the point P.

If 'dB' is a magnetic field at a point due to this small element, it has been observed that

$$dB \propto I, \qquad dB \propto dl, \qquad dB \propto \sin\theta \qquad \text{and} \qquad dB \propto \frac{1}{r^2}$$

Combining all these terms, we get

$$dB \propto \frac{I\, dl\, \sin\theta}{r^2}$$

$$\boxed{dB = K\frac{I\, dl\, \sin\theta}{r^2}} \qquad\qquad \dots (4.5)$$

where K is constant of proportionality and its value depends upon the system of units used and the surrounding medium.

Equation (4.5) is the scalar form of Biot-Savart's law.

Biot-Savart's law can be stated as

The magnetic field at a point due to an element of a conductor carrying a current is directly proportional to :

(i)   the current in the element,

(ii) the length of the element,

(iii) the sine of angle between the element and the line joining the centre of the element to the point, and

(iv) inversely proportional to the square of the distance of the point from the centre of the element.

In S.I. units, value of constant in equation (4.5) is

$$K = \frac{\mu_o}{4\pi} \text{ (for free space)}$$

where μ_o is called the permeability of free space.

$$\mu_o = 4\pi \times 10^{-7} \text{ Wb/A-m}$$

$$\therefore \qquad \frac{\mu_o}{4\pi} = 10^{-7} \text{ Wb/A-m}$$

---

[*]   *When a current flows through an element of length $\vec{dl}$, the quantity $I\,\vec{dl}$ is called current element.*

Equation (4.5) can be written as

$$d\vec{B} = \frac{\mu_o}{4\pi} \frac{I\, dl \sin\theta}{r^2} \text{ tesla} \qquad (\because 1 \text{ tesla} = 1 \text{ Wb/m}^2) \dots (4.6)$$

In Fig. 4.4, according to the rule of cross product, direction of $d\vec{B}$ is perpendicular to the plane containing $\vec{dl}$ and $\vec{r}$ i.e. perpendicular to the plane of the page and pointing inward.

In Fig. 4.4, according to the rule of cross product, direction of $d\vec{B}$ is perpendicular to the plane containing $\vec{dl}$ and $\vec{r}$ i.e. perpendicular to the plane of the page and pointing inward.

In vector form, the equation (4.6) can be written as

$$d\vec{B} = \frac{\mu_o}{4\pi} \frac{I\, \vec{dl} \times \vec{r}}{r^3} \qquad \dots (4.7)$$

The equation (4.7) is the **mathematical form of Biot-Savart's law.**

The total magnetic field can be obtained by integrating equation (4.7) over the whole length.

$$\vec{B} = \int d\vec{B} = \frac{\mu_o}{4\pi} \int_l \frac{I\, \vec{dl} \times \vec{r}}{r^3} \qquad \dots (4.8)$$

Comparison between Coulomb's law and Biot-Savart's law

- Both the electric and magnetic field depend inversely on square of the distance between the source and field point. Both of them are long range forces.
- Charge element dq producing electric field is a scalar, whereas the current element $I\mathbf{dl}$ is a vector quantity having direction same as that of flow of current.
- According to Coulomb's law, the magnitude of electric field at any point P depends only on the distance of the charge element from any point P. According to Biot-Savart's law, the direction of the magnetic field is perpendicular to the current element as well as to the line joining the current element to the point P.
- Both electric field and magnetic field are proportional to the source strength namely charge and current element respectively. This linearity makes it simple to find the field due to more complicated distribution of charge and current by superposing those due to elementary charges and current elements.

Examples (Applications of Biot-Savart's Law)

- The Biot-Savart's law is used for computing the resultant magnetic field **B** at position **r** generated by a *steady* current I (for example magnetic field $\bar{B}$ due to a current carrying wire): a continual flow of charges which is constant in time and the charge neither accumulates nor depletes at any point. It can be used in the calculation of magnetic responses even at the atomic or molecular level, provided that the current density can be obtained from a quantum mechanical calculation or theory.

- The Biot-Savart's law is also used in aerodynamic theory to calculate the velocity induced by vortex lines. Let us study two examples of it.

4.3.2 Long Straight Conductor

Consider a straight wire AB carrying a current I in the direction from A to B. To find the magnetic induction at a point P, consider a small element of wire of length dl (shown in Fig. 4.5). Let the point P be at a distance r from the centre of the element. Let θ be the angle made by small element $(\overrightarrow{dl})$ with line MP. Draw PO perpendicular to the length of wire and let PO = x. According to Biot-Savart's law, the contribution of magnetic field at point P, due to the element $\overrightarrow{dl}$ of the wire is

$$d\overrightarrow{B} = \frac{\mu_o}{4\pi} \frac{I\overrightarrow{dl} \times \overrightarrow{r}}{r^3}$$

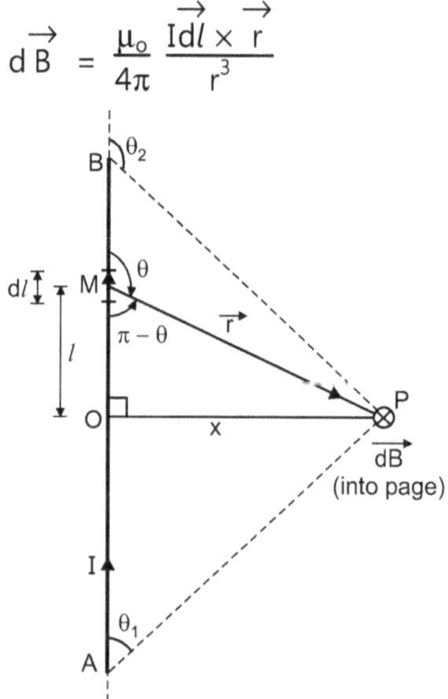

Fig. 4.5 : Magnetic field due to straight wire AB

The direction of $d\vec{B}$ is perpendicular to the plane containing $d\vec{l}$ and $\vec{r}$. So in this case it is perpendicular to the plane of page and pointing in the inward direction.

The magnitude of magnetic induction dB at point P is

$$dB = \frac{\mu_o}{4\pi} \frac{I \, dl \, \sin\theta}{r^2} \qquad \qquad \text{... (4.9)}$$

Let us write r in terms of x and θ.

From Fig. 4.5, $\sin(\pi - \theta) = \dfrac{x}{r}$

$$\sin\theta = \frac{x}{r} \qquad \qquad [\because \sin(\pi - \theta) = \sin\theta]$$

$$\therefore \qquad r = x \operatorname{cosec}\theta \qquad \qquad \text{... (4.10)}$$

Similarly from Fig. 4.5,

$$\cot(\pi - \theta) = \frac{l}{x} \quad \text{or} \quad l = x\cot(\pi - \theta)$$

$$l = -x\cot\theta \qquad \qquad [\because \cot(\pi - \theta) = -\cot\theta]$$

$$\therefore \qquad dl = x \operatorname{cosec}^2\theta \, d\theta \qquad \qquad \text{... (4.11)}$$

Substituting equations (4.10) and (4.11) in equation (4.9), we get

$$dB = \frac{\mu_o I}{4\pi} \frac{x \operatorname{cosec}^2\theta \, d\theta \, \sin\theta}{x^2 \operatorname{cosec}^2\theta}$$

$$dB = \frac{\mu_o I}{4\pi x} \sin\theta \, d\theta$$

As the direction of the magnetic field due to each element is same, the resultant magnetic induction at point P due to whole wire can be obtained by integrating above equation within the limits θ_1 and θ_2.

$$\int_A^B dB = \frac{\mu_o I}{4\pi x} \int_{\theta_1}^{\theta_2} \sin\theta \, d\theta$$

$$B = \frac{\mu_o I}{4\pi x} (-\cos\theta)_{\theta_1}^{\theta_2}$$

or
$$\boxed{B = \frac{\mu_o I}{4\pi x} [\cos\theta_1 - \cos\theta_2]} \qquad \qquad \text{... (4.12)}$$

where θ_1 and θ_2 are values of θ corresponding to the lower end and upper end respectively.

This is the expression for the magnetic field due to wire AB carrying a current I.

Infinitely long wire :

For an infinitely long wire, angles θ_1 and θ_2 tend to 0 and π respectively.

Substituting $\theta_1 = 0$ and $\theta_2 = \pi$ in equation (4.12), we get

$$B = \frac{\mu_o I}{4\pi x}[1 + 1]$$

or
$$\boxed{B = \frac{\mu_o I}{2\pi x}}$$
... (4.13)

This is the expression for a magnetic induction at a point P due to infinitely long wire. The part of the magnetic field around a long, straight, current carrying conductor is as shown in Fig. 4.6.

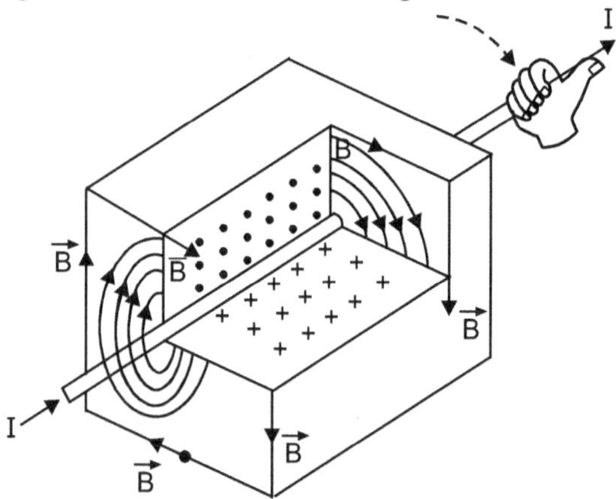

Fig. 4.6 : Magnetic field around a long straight conductor

The field lines are circles with direction determined by right hand rule.

Right hand rule for magnetic field around a current carrying wire : Grasp the wire in your right hand and point the thumb of your right hand in the direction of current. Your fingers will be curled around the wire in the direction of magnetic field.

4.3.3 Magnetic Field due to Circular Current Loop

- In a transformer, an electric motor or an electromagnet, you will find coils of wire with large number of turns. They are closely spaced such that each wire or turn is nearly a planer circular loop. A current in such coil is used to establish a magnetic field. Consider a circular loop of radius 'a' carrying a current I as shown in Fig. 4.7. The expression for magnetic induction at any point on the axis can be obtained by using Biot-Savart's law. Let the point P be situated on the axis of loop at a distance x from centre O of the circular loop. To find the magnetic induction at a point P, consider a small element of circular loop of length dl. Let the point P be at a distance r from the centre of this element.

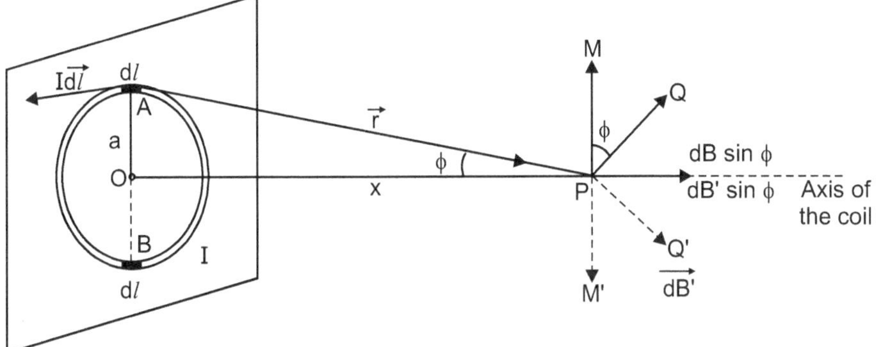

Fig. 4.7 : Magnetic field due to circular current loop

The magnitude of $d\vec{B}$ is $dB = \dfrac{\mu_o}{4\pi}\dfrac{I\,dl\,\sin\theta}{r^2}$

From Fig. 4.7, AP is perpendicular to the small element dl i.e. $\theta = 90°$.

$$\therefore \qquad dB = \frac{\mu_o}{4\pi}\frac{I\,dl}{r^2}$$

The direction of $d\vec{B}$ is perpendicular to the plane containing $\vec{dl}$ and $\vec{r}$ i.e. along $\vec{PQ}$.

The magnetic induction $\vec{dB}$ directed along PQ can be resolved into two components :

(i)   dB cos φ at right angles to the axis (upward shown by PM).

(ii) dB sin φ along the axis of the loop.

Consider another element dl, diametrically opposite, at point B. The magnetic induction $(\vec{dB'})$ produced by this current element will be equal in magnitude but directed along $\vec{PQ'}$. dB' can also be resolved into two components :

(i)   dB' cos φ at right angles to the axis (downward shown by PM').

(ii) dB' sin φ along the axis of the loop.

dB cos ϕ along PM and dB' cos ϕ along PM' cancel each other since they are equal and opposite. However, the component dB sin ϕ due to both elements are in same direction and hence these are added up. This happens for all such pairs of elements. Thus, the resultant magnetic field due to the loop is directed along the axis of the loop and its magnitude

is
$$B = \int dB \sin \phi = \frac{\mu_o I}{4\pi} \int \frac{dl}{r^2} \sin \phi$$

From Fig. 4.7, $\quad r^2 = a^2 + x^2$ and

$$\sin \phi = \frac{a}{r} = \frac{a}{\sqrt{a^2 + x^2}} \qquad \qquad \text{... (4.14)}$$

$$\therefore \qquad B = \frac{\mu_o I}{4\pi} \int \frac{dl}{(a^2 + x^2)} \frac{a}{\sqrt{a^2 + x^2}}$$

$$= \frac{\mu_o I a}{4\pi (a^2 + x^2)^{3/2}} \int dl \qquad \qquad \text{... (4.15)}$$

But $\int dl = 2\pi a$, length of circular loop

$$\therefore \qquad B = \frac{\mu_o I a}{4\pi (a^2 + x^2)^{3/2}} \cdot 2\pi a$$

$$\boxed{B = \frac{\mu_o I a^2}{2 (a^2 + x^2)^{3/2}} \text{ ...(on the axis of circular loop)}} \qquad \text{... (4.16)}$$

This is the expression for the magnetic induction at a point P due to circular loop carrying current I.

If the coil or loop consists of 'N' number of turns of wire, each turn will produce magnetic induction at point P in the same direction. The resultant magnetic induction is

$$B = \frac{\mu_o NI a^2}{2 (a^2 + x^2)^{3/2}} \text{ (on the axis of N circular loops)} \qquad \text{... (4.17)}$$

Special Cases :

(i) Magnetic induction at the centre (O) of the coil or loop :

When the point P is situated at centre (O) of the loop, we have x = 0.

$$\therefore \qquad B = \frac{\mu_o I a^2}{2 (a^2)^{3/2}} \quad \text{or} \quad B = \frac{\mu_o I}{2a} \qquad \text{... (4.18)}$$

Equation (4.18) is the expression for magnetic induction at the centre of current carrying loop.

For N circular loops, $\quad B = \dfrac{\mu_o NI}{2a} \qquad$ ... (at the centre of N circular loops)

(ii) Magnetic induction at a point situated at large distance away from the centre of the coil : If the point P is situated at a large distance away on the axis of coil, we have x >> a.

For x >> a, equation (4.17) can be written as

$$B = \frac{\mu_o \, I \, a^2}{2 \, (x^2)^{3/2}}$$

or

$$\boxed{B = \frac{\mu_o \, I \, a^2}{2x^3}}$$

... (4.19)

This is the expression for magnetic induction at a point P (on the axis) situated at large distance away from the coil.

4.4 Ampere's Circuital Law

- In previous section, we have studied Biot-Savart's law and its application to find the magnetic field produced by a given distribution of currents. Ampere's law is an alternative and equivalent to Biot-Savart's law. Ampere's circuital law gives another method to calculate magnetic field that is produced by a given distribution of current.

Andrew Ampere (1775-1836) was a French physicist, mathematician and chemist who founded the science of electrodynamics. He grasped the significance of Oersted's discovery and carried out several large series of experiments to explore the relation between current electricity and magnetism.

Andrew Ampere

4.4.1 Statement

Ampere's circuital law states that, **the line integral of magnetic field $\vec{B}$, around a closed curve (or loop) is equal to μ_o times the total current I through the area bounded by the curve.**

i.e.,

$$\oint \vec{B} \cdot d\vec{l} = \mu_o I$$

... (4.20)

The circle on the integral sign means that scalar (or dot) product $\vec{B} \cdot d\vec{l}$ is to be integrated around a closed loop, called an **Amperian loop**.

Proof : For the sake of simplicity, consider an infinitely long straight conductor (or wire) carrying a steady current I and the closed curve (C) being a circle in a plane perpendicular to the conductor. The magnetic field lines due to such current carrying conductor are concentric circular loops around it. The sense of this magnetic induction B is anticlockwise.

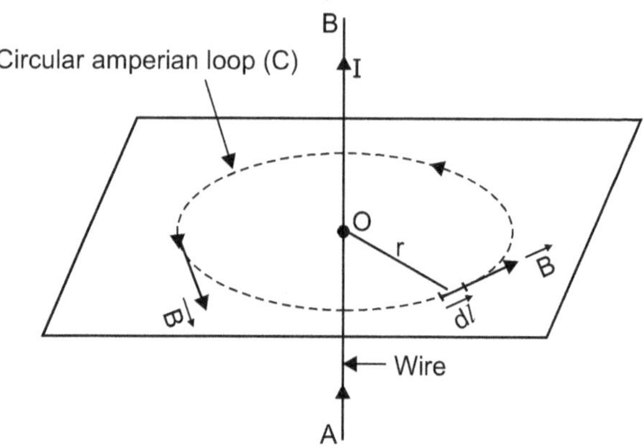

Fig. 4.8 : Ampere's law

Fig. 4.8 shows a small part of infinitely long straight wire in which current flows from A to B. The wire is kept in the free space. Since every point on the circular Amperian loop is at the same distance from the centre of the conductor, the magnetic induction $\vec{B}$ has same magnitude at every point on the loop.

According to Biot-Savart's law, the magnitude of magnetic induction at any point on the loop is

$$B = \frac{\mu_o I}{2\pi r} \qquad \text{... (4.21)}$$

where r is the distance of the point from wire or conductor.

The line integral of $\vec{B}$ around the chosen loop is $\oint_C \vec{B} \cdot \vec{dl}$

If we integrate *counterclockwise*, $\vec{dl}$ and $\vec{B}$ at any point on the loop are parallel to each other i.e. $\theta = 0$.

$$\therefore \quad \oint_C \vec{B} \cdot \vec{dl} = \oint_C B \, dl \cos\theta = \oint_C B \, dl \quad (\because \cos 0 = 1)$$

$$= B \oint_C dl \quad (B \text{ is constant at any point on the loop})$$

$$= 2\pi r B \qquad \left[\because \int dl = \left(\begin{array}{c}\text{circumference} \\ \text{of the loop}\end{array}\right) = 2\pi r\right]$$

Using equation (4.21), we can write

$$\oint_C \vec{B} \cdot \vec{dl} = 2\pi r \cdot \frac{\mu_o I}{2\pi r}$$

$$\therefore \quad \boxed{\oint_C \vec{B} \cdot \vec{dl} = \mu_o I}$$

The above relation is **Ampere's circuital law**.

Importance of Ampere's Law

1. Ampere's law may be derived from the Biot-Savart's law and Biot-Savart's law can also be derived from the Ampere's law. Thus, the two laws are equivalent in scientific content. It's relationship to the Biot-Savart's law is similar to the relationship between Gauss and Coulomb law. However, Ampere's law is more conveniently used to symmetrical current distribution.

2. Use of Biot-Savart's law to determine $\vec{B}$ due to a given current distribution involves integrals which are quite complicated to evaluate. Use of Ampere's law to such situations eliminates complicated integral.

3. When the closed loop encloses number of current carrying conductors, then $\vec{B}$ is the resultant of magnetic fields due to all of them and I is the total current through a surface enclosed by a loop. *This is the principle of superposition.*

4. Ampere's law is true for steady current which does not fluctuate with time.

5. The circulation of $\overrightarrow{B}$ over the closed loop which does not enclose the current carrying conductor is zero, because under this condition, the R.H.S. of equation (4.21) is zero. Ampere's circuital law is used to determine magnetic induction for current distribution with sufficient symmetry (like Gauss's law in electrostatics used to determine electric intensity).

Applications of Ampere's Circuital law :

4.4.2 Field of a Solenoid

A solenoid consists of a long wire in the form of helical winding on a cylinder usually circular in cross-section. The wire is coated with insulating material so that although the adjacent turns physically touch each other, they are electrically insulated. Generally, the length of the solenoid is large as compared to transverse direction. Solenoids are very common electrical devices that are important in many technological applications to obtain uniform magnetic field.

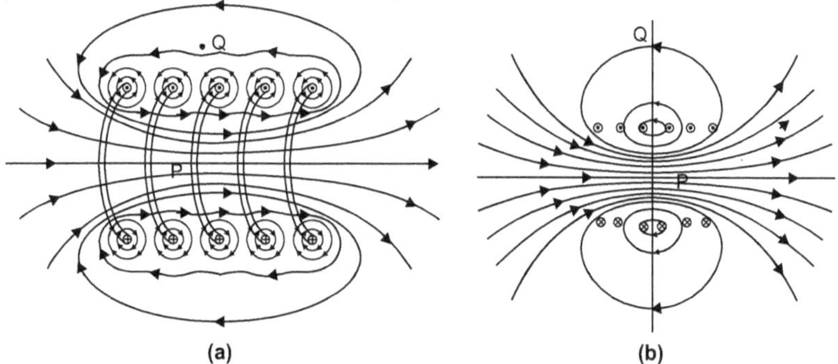

(a)   (b)

Fig. 4.9 : (a) The magnetic field due to a section of the solenoid which has been stretched out for clarity. Only the interior semi-circular part is shown. Notice how the circular loops between neighbouring turns tend to cancel. (b) The magnetic field of a finite solenoid

Fig. 4.9 displays the magnetic field lines for finite solenoid. In Fig. 4.9 (a), we have shown a section of solenoid in enlarged manner. From this Fig. 4.9 (a), it is clear from the circular loops that the field lines tend to cancel between two neighbouring turns. Fig. 4.9 (b) shows entire finite solenoid with magnetic field. Also it is noticed that field at the interior midpoint P is uniform and strong. The field at the exterior mid-point Q is weak and along the axis of solenoid with no perpendicular or normal component.

As solenoid appears like a long cylindrical metal sheet, the field outside the solenoid approaches zero. The upper views of dots in Fig. 4.10 are like a uniform current sheet coming out of the plane of paper and lower row of crosses is like a uniform current sheet going into the plane of the paper.

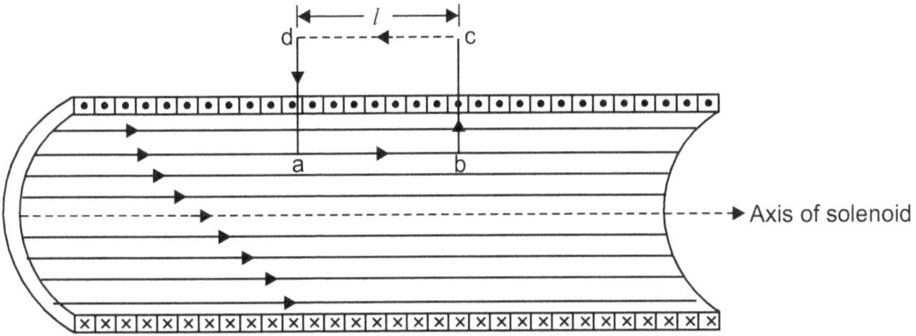

Fig. 4.10 : Section of a long solenoid carrying a current I.

The Amperian loop is the rectangle abcda

Let us consider a solenoid having length very large compared with its diameter. It is carrying a current I_o. Fig. 4.10 shows a vertical cross-section through the axis of a solenoid. For application of Ampere's law, consider a rectangular loop abcda such that side ab is along the axis of a solenoid and side cd parallel to axis.

Ampere's circuital law is $\oint \vec{B} \cdot \vec{dl} = \mu_o I$

The integral $\oint \vec{B} \cdot \vec{dl}$ can be written as the sum of four integrals, one for each path segment.

$$\therefore \quad \oint_{abcda} \vec{B} \cdot \vec{dl} = \int_a^b \vec{B} \cdot \vec{dl} + \int_b^c \vec{B} \cdot \vec{dl} + \int_c^d \vec{B} \cdot \vec{dl} + \int_d^a \vec{B} \cdot \vec{dl} \quad \text{... (4.22)}$$

Now, $\vec{B}$ outside the solenoid is zero, hence

$$\int_c^d \vec{B} \cdot \vec{dl} = 0 \quad \text{... (4.23)}$$

Since $\vec{B}$ inside the solenoid is parallel to the axis of solenoid, path segments bc and da are perpendicular to $\vec{B}$, hence

$$\int_b^c \vec{B} \cdot \vec{dl} = \int_d^a \vec{B} \cdot \vec{dl} = 0$$

$$(\vec{B} \cdot \vec{dl} = Bdl \cos \theta \text{ and } \theta = 90°) \quad \ldots (4.24)$$

Substituting equations (4.23) and (4.24) in equation (4.22), we get

$$\oint_{abcda} \vec{B} \cdot \vec{dl} = \int_a^b \vec{B} \cdot \vec{dl} = \int_a^b Bdl \cos \theta$$

Since $\vec{B}$ and $\vec{dl}$ are parallel at any point on the path ab, we have $\theta = 0$.

$$\therefore \quad \oint_{abcda} \vec{B} \cdot \vec{dl} = \int B \, dl = Bl$$

where ab = length of segment = l

But according to Ampere's law, $\oint \vec{B} \cdot \vec{dl} = \mu_o I$

$$\therefore \quad Bl = \mu_o I \quad \ldots (4.25)$$

where B is magnetic field at a point well inside the solenoid and I is the current crossing the area enclosed by the closed loop. If a given solenoid consists of N turns uniformly wound over length L, number of turns per unit length is $n = \dfrac{N}{L}$.

$$\therefore \quad \text{Number of turns in length } l = \left(\dfrac{N}{L}\right) l = nl$$

Current flowing through the area enclosed by a given loop is

$$I = nl \, I_o \quad \ldots (4.26)$$

Substituting equation (4.26) in equation (4.25), we get

$$Bl = \mu_o \, nl \, I_o \quad \text{or} \quad \boxed{B = \mu_o \, nI_o} \quad \ldots (4.27)$$

Thus, from this equation, it is clear that B is independent of length and diameter of the solenoid and is uniform over the cross-section of the solenoid. This expression (4.27) however applies to a long solenoid in free space and for points well inside the solenoid. A solenoid is a very important coil or wire that is used in inductors, electromagnets, antennas, valves, and many more. The application of a solenoid varies in many different types of industries. It can be used in a simple locking device, medical clamping equipment, an automotive gear box, and an air conditioning unit. Solenoids are highly essential in many automated applications today.

Use of solenoid in magnetic resonance imaging : M.R.I. is a medical imaging technique used in radiology to visualize structure of a body in detail.

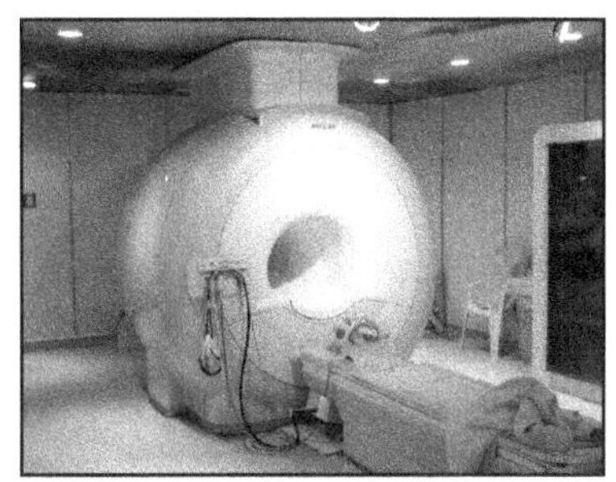

Fig. 4.11

4.4.3 Field of a Toroid

- A toroid is a hollow circular ring on which large number of turns of a wire is closely wound. (The shape of a toroid is like a doughnut or similar to Indian dish *medu wada*). It can be considered as a solenoid which has been bent into circular shape to close on itself. A toroid is used as an inductor in electronic circuits, especially at low frequencies where comparatively large inductances are necessary. Fig. 4.12 shows a toroid carrying a current I_o through its winding.

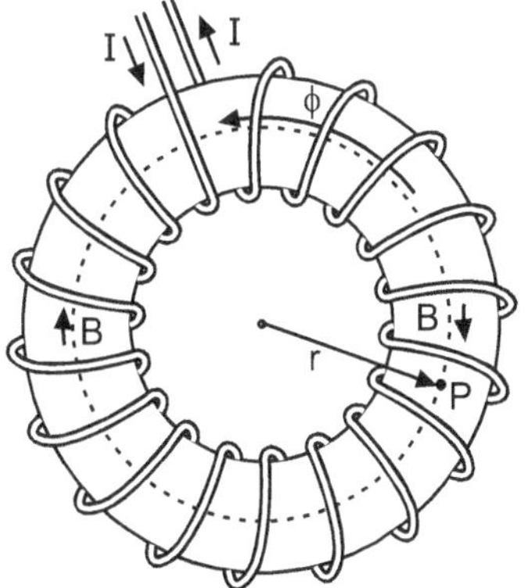

Fig. 4.12 : Toroid

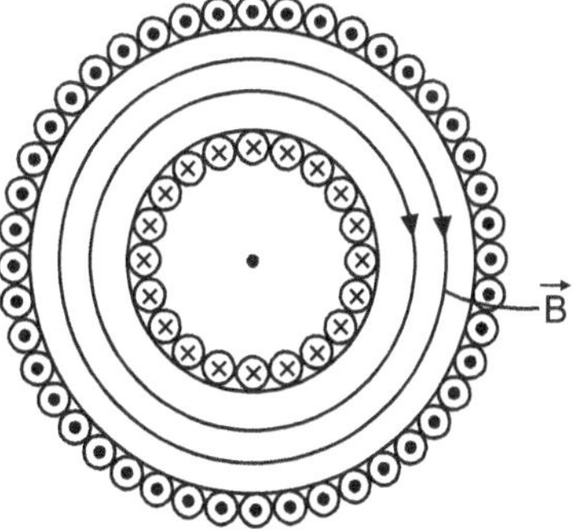

Fig. 4.13 : $\vec{B}$ inside a toroid (horizontal cross-section of toroid)

- We have to find the magnetic induction at a point P inside a toroid. Let the point P be at a distance r from the centre O. To find B at a point P, let us take a concentric circle of radius r as an Amperian loop (shown by dotted line in figure) and traverse it in the clockwise direction.

By symmetry, field lines of $\vec{B}$ form concentric circles (Refer Fig. 4.13) inside the toroid and the field will have equal magnitude at all points on the Amperian loop.

Applying Ampere's law to the chosen loop, we write

$$\oint \vec{B} \cdot \vec{dl} = \mu_o I$$

where I is the current crossing the area enclosed by the chosen loop.

Now, $$\oint \vec{B} \cdot \vec{dl} = \oint B dl \cos \theta$$

$$= B \oint dl \qquad (\because \ \theta = 0 \text{ and } \cos 0 = 1)$$

or $$\oint \vec{B} \cdot \vec{dl} = B\, 2\pi r \qquad \qquad \dots (4.28)$$

If N is the total number of turns in the given toroid, the current crossing the area bounded by the circle is NI_o.

$\therefore \qquad\qquad I = NI_o \qquad\qquad \dots (4.29)$

Substituting equations (4.28) and (4.29) in equation (4.20), we get

$$B\, 2\pi r = \mu_o NI_o$$

or $$B = \frac{\mu_o NI_o}{2\pi r} \qquad \text{(toroidal solenoid)} \dots (4.30)$$

If l is the mean circumference of the toroid then $l = 2\pi r$.

$$\therefore \qquad B = \frac{\mu_o NI_o}{l}$$

or $$\boxed{B = \mu_o\, nI_o} \quad \text{where } n = N/l.$$

This is the expression for magnetic field inside the toroid.

Note : If we denote current passing through the winding of toroid by I (instead of I_o), we get

$$B = \frac{\mu_o NI}{l} = \mu_o nI \qquad\qquad \dots (4.31)$$

- From the equation (4.31), it is clear that B is not constant at different points over the cross-section of the toroid core because r is larger on the outer side of the section than the inner side of toroid. In an ideal toroid, the coils are circular. But in reality, the turns of toroid coil form a helix and there is always a small magnetic field external to the toroid.

- Toroids are found in thermonuclear fusion research to maintain constant control of nuclear reactions. Toroids are expected to play very important role in an equipment **tokamak** for plasma confinement in fusion power reactors. For most of the practical purposes, they can be used for windings in air or on a core of any non-magnetic, non-superconducting material.

4.5 Magnetic Flux and Gauss's Law for Magnetism

Karl Friedrich Gauss was child prodigy and gifted in mathematics, physics, engineering, astronomy and even land surveying. The properties of numbers fascinated him and in his work he anticipated major mathematical development of later times. His mathematical theory of curved surface laid the foundation for later work of Riemann.

Karl Friedrich Gauss
(1777-1855)

Magnetic Flux

We know that the electric flux $d\phi$ through an area element $d\vec{S}$ is defined as

$$d\phi = \vec{E} \cdot d\vec{S} = E\, dS \cos\theta$$

We can define the magnetic flux ϕ_B through a surface as we have defined electric flux.

We can divide any surface into elements of area dA. (Refer Fig. 4.14) $\vec{B}$ can be resolved into $B_\perp$ (perpendicular) component and $B_{||}$ (parallel) component. For each element, we determine $B_\perp$, the component of $\vec{B}$ normal to the surface at position as shown in Fig. 4.14.

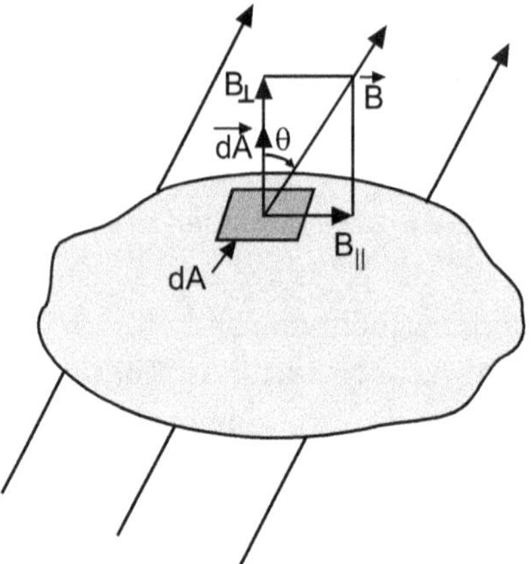

Fig. 4.14 : Magnetic flux through an area element dA

From Fig. 4.14, $B_\perp$ = B cos θ

where θ is the angle between $\overrightarrow{B}$ and line perpendicular to the surface.

We define magnetic flux $d\phi_B$ through this area as

$$d\phi_B = B_\perp \, dA = B \cos\theta \, dA$$

$$d\phi_B = \overrightarrow{B} \cdot d\overrightarrow{A} \qquad \text{... (4.32)}$$

The total magnetic flux through the surface is the sum of the contribution from the individual area elements.

$$\phi_B = \int B_\perp \, dA = \text{Magnetic flux through surface}$$

$$\phi_B = \int B \cos\theta \, dA$$

$$= \int \overrightarrow{B} \cdot d\overrightarrow{A} \quad \text{(magnetic flux through a surface) ... (4.33)}$$

Magnetic flux is a scalar quantity. When $\overrightarrow{B}$ is perpendicular to the surface, then cos θ = 1.

$$\therefore \qquad \phi_B = B\,A \qquad \text{... (4.34)}$$

Magnetic flux is expressed in Weber.

The unit is named as Weber in the honour of German physicist Wilhelm Weber (1804-1891).

$$\therefore \qquad 1\,\text{Wb} = 1\,\text{T} \cdot \text{m}^2$$

One Weber (1 Wb) is equal to the unit of magnetic field (1 T) times the unit of area (1 m^2).

In Gauss law (in electrostatics) the total electric flux through a closed surface is proportional to the total electric charge enclosed by surface. For problem, if the closed surface encloses an electric dipole, the total electric flux is zero because the total charge is zero.

Similarly, if there were such things as a single magnetic charge (magnetic monopole), the total magnetic flux through a closed surface would be proportional to the total magnetic charge enclosed. But we know that no magnetic monopole has ever been observed. Hence we conclude that, total magnetic flux through a closed surface is zero.

Gauss's law for magnetism is stated as the total magnetic flux through a closed surface is always zero.

- Symbolically, $\phi_B = \oint \vec{B} \cdot d\vec{A} = 0$

$$\text{(Gauss's law for magnetic field) ... (4.35)}$$

- This is in contrast with Gauss's law for electric field because

$$\phi_{ele} = \oint \vec{E} \cdot d\vec{A} = \frac{q_{enc}}{\varepsilon_o} \qquad \text{(Gauss's law for electric field)}$$

In both the equations, the integral is taken over a closed surface. Gauss's law for electric field says that this integral (the net electric flux at the surface) is proportional to the net electric charge q_{enc} enclosed by the surface. But Gauss's law for magnetic fields says that there cannot be net magnetic flux at the surface. This is because there can be no net magnetic charge (individual magnetic poles) enclosed by the surface. The simplex magnetic structure that can exist and thus enclosed by Gaussian surface is a dipole, which consists of both a source and a sink for the field lines. Thus, their magnetic flux into the surface and out of the surface is same. Hence, net magnetic flux must be zero. Gauss's law for magnetic fields holds for more complicated structure than a magnetic dipole and it holds even if the Gaussian surface does not enclose the entire structure. Gaussian surface II near the bar magnet of Fig. 4.15 encloses no poles. We can conclude that the net magnetic flux at it is zero. Gaussian surface I is more difficult to understand. It seems that it may enclose only the label N of magnet and not label S. However, a south pole must be associated with lower boundary of surface because magnetic field lines enter the surface there. The enclosed surface is like broken magnets where N and S poles cannot be separated i.e. magnetic monopole does not exist.

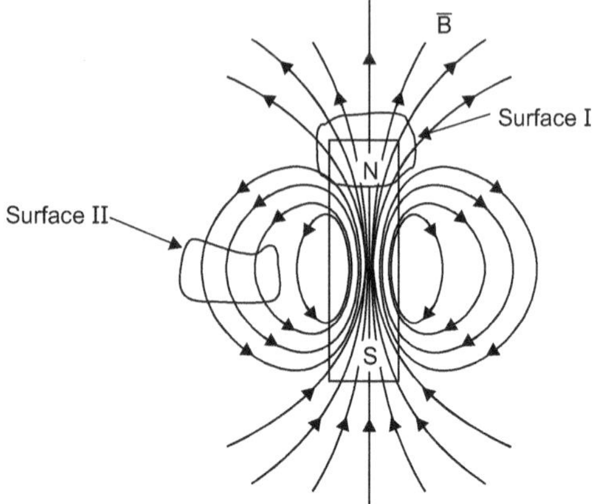

Fig. 4.15

- Thus Gaussian surface I encloses a magnetic dipole and the net flux through the surface is zero. If the element dA in equation (4.35) is at right angle to field lines, then $B_\perp = B$.

$$B = \frac{d\phi_B}{dA_\perp} \qquad \dots (4.36)$$

i.e. magnitude of magnetic field is equal to flux per unit area across an area at right angle to the magnetic field. Hence, the magnetic field $\vec{B}$ is sometimes called *magnetic flux density*.

Solved Problems

Problem 4.1 : *A proton is projected with speed of 6×10^6 m/s horizontally from east to west. A uniform magnetic field $\vec{B}$ of strength 4×10^{-3} T exists in vertically upward direction. Find the force on proton just after its projection. Also calculate the acceleration produced.*

(Given : mass of proton = 1.67×10^{-27} kg)

Solution :

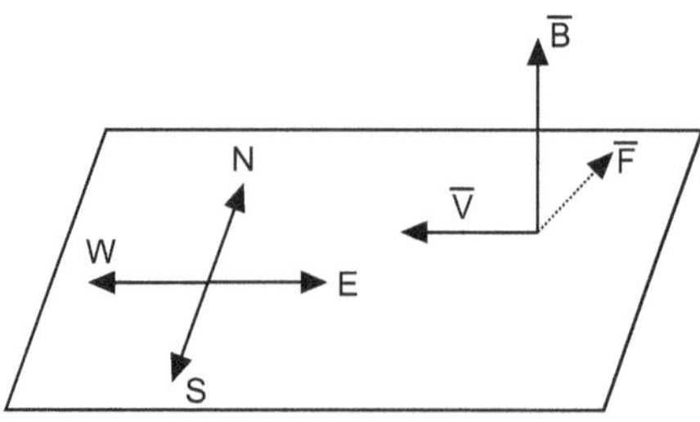

Fig. 4.16

As shown in Fig. 4.16, the force is perpendicular to $\vec{B}$, hence it is in horizontal plane through the proton. In this plane, it is perpendicular to the velocity $\vec{v}$. Thus, it is along north-south line. Thus, $\vec{v} \times \vec{B}$ is towards north. As charge on proton is positive, the force $\vec{F} = q\vec{v} \times \vec{B}$ is also towards north.

The magnitude of force is

$$F = qvB = (1.6 \times 10^{-19} \text{ C})(6.00 \times 10^6 \text{ m/s})(4 \times 10^{-3} \text{ T})$$
$$= 3.84 \times 10^{-15} \text{ N}$$

The acceleration of proton,

$$a = \frac{F}{m} = \frac{3.84 \times 10^{-15} \text{ N}}{1.67 \times 10^{-27} \text{ kg}} = \mathbf{2.3 \times 10^{12} \text{ m/s}^2} \qquad \textbf{... Ans.}$$

Problem 4.2 : *An aluminium wire of diameter 0.4 cm carries a current of 25 ampere. Find the magnetic field at the surface of the wire.*

(April 16, 11; Oct. 10)

Solution : The magnetic field at a point distant r from a straight current carrying wire is given by

$$B = \frac{\mu_o I}{2\pi r}$$

where I is the current flowing through the wire.

At the surface of the wire r = R, radius of wire.

$$B = \frac{\mu_o}{2\pi} \frac{I}{R}$$

$$\mu_o = 4\pi \times 10^{-7} \text{ Wb/A-m}$$

$$\text{Diameter} = 0.4 \text{ cm}$$

$$\therefore \qquad \text{Radius R} = 0.2 \text{ cm} = 0.2 \times 10^{-2} \text{ meter}$$

$$\therefore \qquad B = \frac{4\pi \times 10^{-7}}{2\pi} \times \frac{25}{0.2 \times 10^{-2}}$$

$$B = \mathbf{25 \times 10^{-4} \text{ Wb/m}^2} \qquad \textbf{... Ans.}$$

Problem 4.3 : *Fig. 4.17 shows two long straight wires carrying electric currents 10 A in opposite directions. The separation between the wires is 6 cm. Find magnetic field at the point P midway between the wires.*

(Oct. 15)

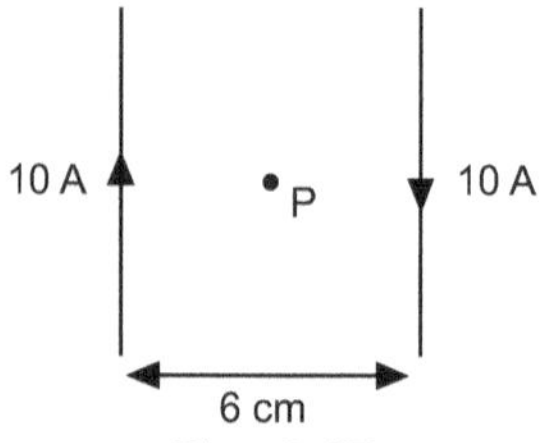

Fig. 4.17

Solution : d = 6 cm, hence x = 3 cm = 0.03 m.

Magnetic field due to each wire is

$$B = \frac{\mu_o\, i}{2\pi x} = \frac{4\pi \times 10^{-7} \times 10}{2\pi \times 3.0 \times 10^{-2}\ m} = \frac{2 \times 10^{-7} \times 10}{3 \times 10^{-2}}$$

$$B = \mathbf{6.67 \times 10^{-5}\ T} \qquad \text{... Ans.}$$

The net field due to both the wires is 13.34×10^{-5}

$$= \mathbf{1.334 \times 10^{-4}\ T} \qquad \text{... Ans.}$$

Problem 4.4 : *Fig. 4.18 shows a view of flat surface with area 4.0 cm^2 in uniform magnetic field. The magnetic flux through this area is 0.707 $\times 10^{-3}$ Wb. Calculate magnitude of the magnetic field.*

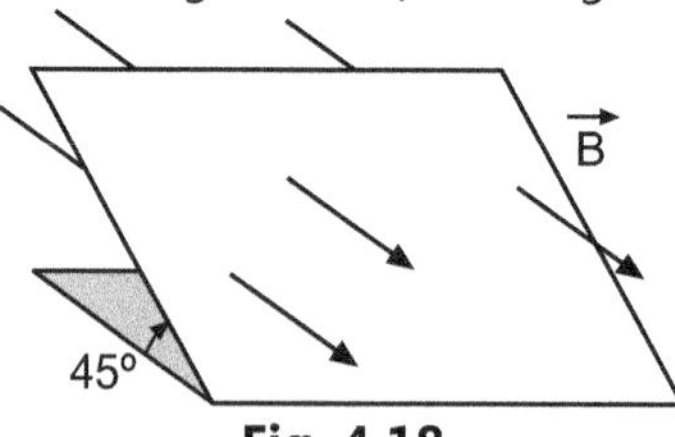

Fig. 4.18

Solution : The area vector A makes angle 45° with B.

∴ θ would have been 135° and magnetic flux would have been negative.

But ϕ_B, B and A all are positive.

∴ cos θ must be positive, so θ = 45°.

$$\therefore \qquad B = \frac{\phi_B}{A \cos \theta} = \frac{0.707 \times 10^{-3}\ Wb}{(4 \times 10^{-4}\ m^2)\,(\cos 45°)} = \mathbf{2.5\ T} \qquad \text{... Ans.}$$

Problem 4.5 : *A charge of 3.0 μC moves with speed 3.0 $\times 10^6$ m/s along positive X-axis. A magnetic field of strength (0.10 $\vec{j}$ + 0.20 $\vec{k}$) T exists in space. Find the magnetic force acting on the charge.* **(Oct. 15)**

Solution : Force on the charge

$$= q\,\vec{v} \times \vec{B} = (3 \times 10^{-6})\,(3 \times 10^6\ \hat{i}) \times (0.10\ \hat{j} + 0.20\ \hat{k})\ T$$

$$= 9\,(0.10\ \hat{i} \times \hat{j} + 0.20\ \hat{i} \times \hat{k}) = \mathbf{(0.9\ \hat{k} - 1.8\ \hat{j})\ N} \qquad \text{... Ans.}$$

Problem 4.6 : *A long wire having a semi-circular loop of radius 'r' carries a current I as shown in Fig. 4.19. Find the magnetic induction at centre O due to entire wire.*

Solution : According to Biot-Savart's law, the magnetic induction at a point due to current element $I \, \overrightarrow{dl}$ is $\overrightarrow{dB} = \dfrac{\mu_o}{4\pi} \dfrac{I \, \overrightarrow{dl} \times \overrightarrow{r}}{r^3}$

where $\overrightarrow{r}$ is radius vector from the element to the point.

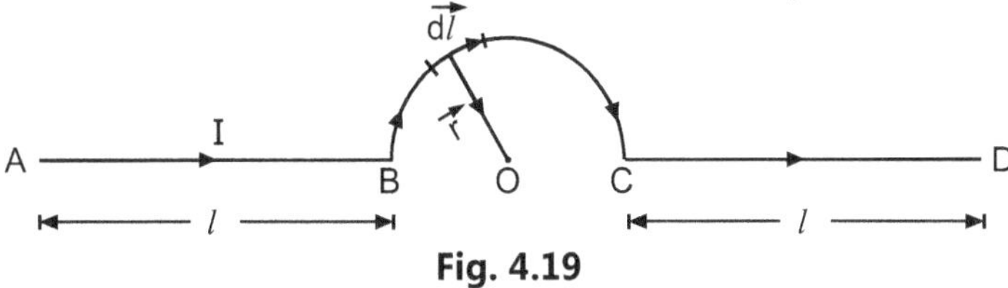

Fig. 4.19

The angle between radius vector $\overrightarrow{r}$ and straight portion AB is $0°$, so $I\overrightarrow{dl} \times \overrightarrow{r}$ is zero.

Also the angle between radius vector $\overrightarrow{r}$ and straight portion CD is $180°$. So $I\overrightarrow{dl} \times \overrightarrow{r}$ is zero.

So there is contribution from semi-circular loop only.

Consider a current element $I\overrightarrow{dl}$. The angle between $I\overrightarrow{dl}$ and radius vector $\overrightarrow{r}$ is $\dfrac{\pi}{2}$.

$$\therefore \quad dB = \frac{\mu_o}{4\pi} \frac{Idl \sin \frac{\pi}{2}}{r^2} = \frac{\mu_o}{4\pi} \frac{Idl}{r^2}$$

The magnetic induction due to whole semi-circular loop is

$$B = \int dB = \frac{\mu_o I}{4\pi r^2} \int dl$$

$$\int dl = \pi r, \text{ length of semi-circular loop}$$

$$\therefore \quad B = \frac{\mu_o I}{4\pi r^2} \pi r$$

$$\mathbf{B} = \frac{\mathbf{\mu_o I}}{\mathbf{4r}} \qquad\qquad \text{... Ans.}$$

This is the magnetic induction at O due to entire wire.

Problem 4.7 : *A coil made of 200 circular loops with radius 0.60 m carries an electric current of 2.5 A. Find the magnetic field at a point along the axis of coil which is at a distance 0.80 m from the centre.*

Solution : $N = 200$, $I = 2.5$ A, $a = 0.60$ m, $x = 0.80$ m.

$$B = \frac{\mu_o \, NI \, a^2}{2 \, (x^2 + a^2)^{3/2}} = \frac{4\pi \times 10^{-7} \times 200 \times 2.5 \times (0.60)^2}{2 \, [(0.80)^2 + (0.60)^2]^{3/2}}$$

$$B = \mathbf{1.1 \times 10^{-4} \ T} \qquad\qquad \textbf{... Ans.}$$

Problem 4.8 : *A closed curve encircles several conductors. The line integral $\oint \vec{B} \cdot \vec{dl}$ around the curve is 3.83×10^{-4} T.m. (a) What is the net current in the conductor ? (b) If you were to integrate around the curve in opposite direction, what would be the value of line integral ?* **(Nov. 10)**

Solution : (a) $\oint \vec{B} \cdot \vec{dl} = \mu_o I$

$$I = \frac{\oint \vec{B} \cdot \vec{dl}}{\mu_o} = \frac{3.83 \times 10^{-4}}{4\pi \times 10^{-7}} = \mathbf{305 \ A} \qquad \textbf{... Ans.}$$

(b) In opposite direction, B will be negative.

$$\therefore \qquad \oint B \, dl = \mathbf{-\ 3.83 \times 10^{-4} \ T \ m} \qquad\qquad \textbf{... Ans.}$$

Problem 4.9 : *A solenoid of length 100 cm is wound uniformly with 10,000 turns of wire. It carries a current of 4 amp. What is the value of (i) magnetic field on the axis of the solenoid at the centre, (ii) magnetic field on the axis at an end ?*

Solution : Given : Length of solenoid (L) = 100 cm = 1 m

Total number of turns (N) = 10,000.

Current passing through wire of solenoid (I) = 4 A.

(i)   The magnitude of magnetic field B at a point situated well inside a solenoid (or at the centre) is given by

$$B = \mu_o \, nI$$

where
$$n = \frac{N}{L} = \frac{10000}{1}$$

$$\therefore \qquad B = 4\pi \times 10^{-7} \times \frac{10000}{1} \times 4 = \mathbf{0.0502 \ Wb/m^2} \qquad \textbf{... Ans.}$$

(ii) The magnetic field at the end is half of that at the centre.

Therefore, magnetic field on the axis at an end $= \dfrac{B}{2} = \mathbf{0.0251 \ Wb/m^2}$.

$$\textbf{... Ans.}$$

Problem 4.10 : *A solenoid of length 0.5 m has a radius of 1 cm and is made of 500 turns. It carries a current of 3 A. What is the magnitude of magnetic field inside the solenoid ?*

Solution : Given : N = 500, I = 3 A, radius = 0.01 m

Number of turns per unit length = n = 500 / 0.5 = 1000 turns/length

r = 0.01 m and l = 0.5 m. Thus, $l/a \sim 50$, $l >> a$

We can use long solenoid formula

$$B = \mu_o nI = 4\pi \times 10^{-7} \times 10^3 \times 3 = \mathbf{3.77 \times 10^{-3}\ T} \qquad \textbf{... Ans.}$$

Problem 4.11 : *A coil of 20 cm radius has 15 turns and carries a current of 3 ampere. Find the magnetic field at the centre of the coil.*

(Given : $\mu_o = 4\pi \times 10^{-7}\ Wb\ A^{-1}\ m^{-1}$).

Solution : Given : Radius of the coil, a = 20 cm = 20×10^{-2} m.

Current through the wire of coil, I = 3 A.

Number of turns N = 15.

Magnetic field at the centre of coil is given by

$$B = \frac{\mu_o NI}{2a} = \frac{4\pi \times 10^{-7} \times 15 \times 3}{2 \times 20 \times 10^{-2}}$$

or

$$B = \frac{2\pi \times 10^{-7} \times 45}{20 \times 10^{-2}}$$

$$B = \mathbf{1.4 \times 10^{-4}\ Wb/m^2} \qquad \textbf{... Ans.}$$

Problem 4.12 : *Two long wires a and b carrying equal current of 10.0 A, are placed parallel to each other with a separation of 4 cm between them as shown in Fig. 4.20.*

Find the magnetic field B at points P, Q and R.

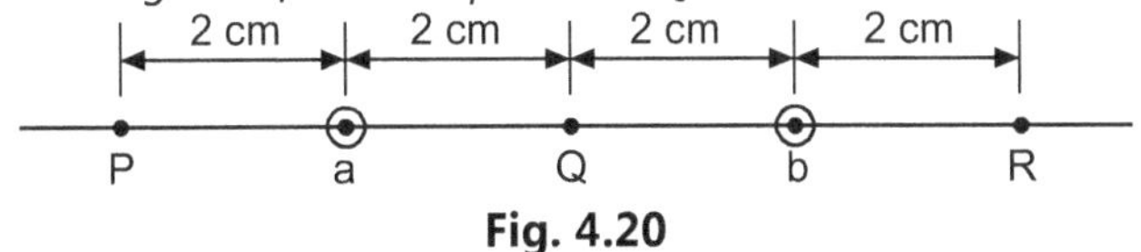

Fig. 4.20

Solution : The magnetic field at P due to wire a has magnitude

$$B_1 = \frac{\mu_o\, i}{2\pi x} = \frac{4\pi \times 10^7 \times \dfrac{Tm}{A} \times 10\ A}{2\pi \times 2 \times 10^{-2}\ m} = 1.00 \times 10^{-4}\ T$$

The direction will be perpendicular to the line shown and point downward in Fig. 4.20.

The field at P due to other wire has magnitude

$$B_2 = \frac{\mu_o i}{2\pi x} = \frac{4\pi \times 10^{-7} \times 10\ A}{2\pi \times 6 \times 10^{-2}\ m} = 0.33 \times 10^{-4}\ T$$

The direction of B_2 will be same as B_1.

$\therefore$ Radius field, $B = B_1 + B_2 = \mathbf{1.33 \times 10^{-4}\ T}$ ... **Ans.**

Similarly, resultant field at R will be 1.33×10^{-4} T. The magnetic field at point Q due to two wires will have equal magnitudes but opposite direction and hence resultant field is zero.

Problem 4.13 : *Using Ampere's law, find the magnetic field at a point inside and outside the long straight cylindrical current carrying wire.*

Solution : Consider a long, straight cylindrical wire carrying current I. Let 'a' be the radius of cylindrical wire. We have to find the magnetic induction at a point inside and outside the conductor using Ampere's circuital law.

(a) Magnetic field at a point situated outside the cylindrical wire : Consider a point P at a perpendicular distance r from the axis of cylindrical wire (Refer Fig. 4.21).

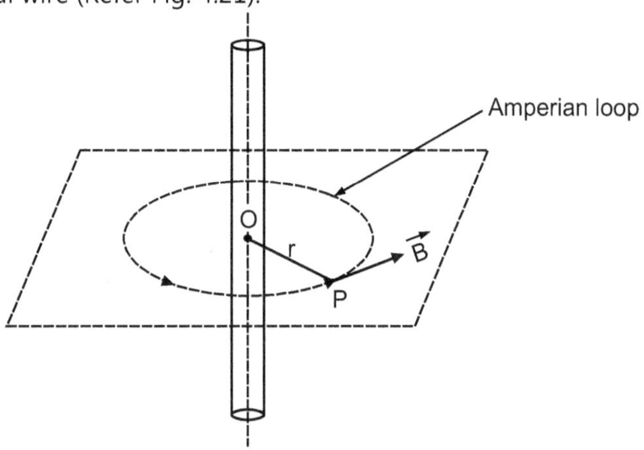

Fig. 4.21

To find magnetic field at a point P, let us draw a closed circular loop (Amperian loop) passing through point P and having centre O.

According to Ampere's law, $\oint \vec{B} \cdot d\vec{l} = \mu_o I_{enc}$

Since I is current passing through the closed loop, $I_{enc} = I$.

or $\oint Bdl \cos\theta = \mu_o I$

At any point on the loop, $\vec{B}$ and $d\vec{l}$ are in same direction, so we get $\theta = 0°$.

$\therefore$ $\oint Bdl = \mu_o I$

Since all points on the circular loop are at same distance from the axis of wire, the magnitude of B at all points is same (i.e. B is constant).

$\therefore$ $B \oint dl = \mu_o I$ or $B\, 2\pi r = \mu_o I$

$\therefore$ $$\boxed{B = \frac{\mu_o I}{2\pi r}}$$ ... **Ans.**

(b) Magnetic field at a point inside the cylindrical wire : Consider a point P' situated inside the conductor at a perpendicular distance R from the axis of cylindrical wire. Fig. 4.22 shows a cross-sectional view of wire.

To find B at point P', let us draw a closed circular loop (Amperian loop) passing through point P and having centre O (shown by dotted line).

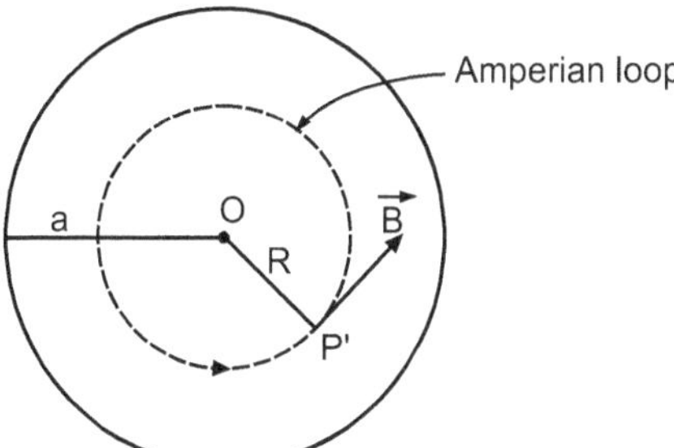

Fig. 4.22 : Cross-section of a wire

According to Ampere's law, $\oint \vec{B} \cdot d\vec{l} = \mu_o I_1$

where I_1 is the current passing through the loop.

Since the current is uniformly distributed, current passing per unit area is same.

i.e. $$\frac{I}{\pi a^2} = \frac{I_1}{\pi R^2}$$

or $$I_1 = \pi R^2 \frac{I}{\pi a^2} = I \frac{R^2}{a^2}$$

$\therefore$ $$\oint \vec{B} \cdot d\vec{l} = \mu_o I \frac{R^2}{a^2}$$

or $$B\, 2\pi R = \frac{\mu_o I R^2}{a^2}$$

$\therefore$ $$\boxed{B = \left(\frac{\mu_o I}{2\pi a^2}\right) R}$$ ... **Ans.**

Think Over It

1. What happens to the direction of the magnetic field about a straight current carrying conductor when the direction of current is reversed ?
2. A magnetic field can exert a force on a moving charged particle, but it cannot change the particle's kinetic energy. Why not?

Summary

- We can represent magnetic field by magnetic lines and a line at any point is tangent to magnetic field.
- We define magnetic flux $d\phi_B$ as

$$d\phi_B = \vec{B} \cdot d\vec{A}$$

$$\therefore \qquad \phi_B = \int \vec{B} \cdot d\vec{A}$$

- The force exerted by a magnetic field is called magnetic force. It is given by equation $\vec{F} = q\vec{v} \times \vec{B}$

- Magnetic field due to current element is given by Biot-Savart's law. Biot-Savart's law is written as $dB = \dfrac{\mu_o}{4\pi} \dfrac{I \, dl \sin\theta}{r^2}$

- Magnetic field due to long straight wire is given by the relation

$$B = \frac{\mu_o I}{2\pi x}$$

  where I is the current flowing through the wire and x is the distance.

- Magnetic field due to circular current loop is

$$B = \frac{\mu_o I a^2}{2 (a^2 + x^2)^{3/2}}$$

- Ampere's circuital law states that, the line integral of magnetic field around a closed curve is μ_o times total current I through the area bounded by the curve.

$$\oint \vec{B} \cdot d\vec{l} = \mu_o I$$

- Magnetic field at a point inside the solenoid is $B = \mu_o n \, I_o$, where n are number of turns per unit length i.e. B is independent of length and diameter of solenoid and is uniform over the cross-section of solenoid.

- Toroid may be considered as a solenoid of finite length bent into shape of circular ring. Magnetic field due to toroid,

$$B = \frac{\mu_o N I_o}{l} = \mu_o I_o \, n \text{ where } n = N/l$$

- Gauss's law for magnetism is stated as the total magnetic flux through a closed surface is zero.

$$\phi_B = \oint \vec{B} \cdot d\vec{A} = 0$$

Exercises

(A) Multiple Choice Questions :

1. In Ampere's circuital law, what is the purpose of an 'Amperian Path'?
   - (a) computation of magnetic field intensity
   - (b) determination of differential element of path length
   - (c) estimation of electric flux density
   - (d) detection of loop in a constant plane

2. What is the direction of magnetic field intensity vector due to infinite long straight filament?
   - (a) radial
   - (b) elliptical
   - (c) parallel to conductor
   - (d) circumferential

3. According to Biot-Savart's law, which parameter exhibits an inverse relationship to the differential magnetic field (dB)?
   - (a) current
   - (b) magnitude of differential length
   - (c) sine of angle between length element and line connecting differential length to point
   - (d) square of the distance from differential element to point

4. Quantity that is not affected by magnetic field is ......
   - (a) moving charge
   - (b) change in magnetic flux
   - (c) current flowing in conductor
   - (d) stationary charge

5. Biot-Savart's law in magnetism is analogous to which law in electricity?
   - (a) Gauss's law
   - (b) Faraday's law
   - (c) Coulomb's law
   - (d) Ampere's law

6. Which of the following cannot be computed using the Biot-Savart's law ?
   - (a) Magnetic field intensity
   - (b) Magnetic flux density
   - (c) Electric field intensity
   - (d) Permeability

7. What will be the magnetic field intensity at a point on the centre of the circular conductor of radius 2 m with current 8 A ?
    (a) 1 unit   (b) 2 units
    (c) 3 units   (d) 4 units
8. What happens to the magnetic field at the centre of circular coil when its radius is doubled keeping current constant ?
    (a) remains same   (b) halved
    (c) doubled   (d) quadrupled

Answers

1. (a)	2. (d)	3. (d)	4. (a)	5. (c)	6. (c)	7. (b)	8. (b)

(B)   Short Answer Type Questions :

1. What do you mean by magnetic field ?
2. State Biot-Savart's law.
3. What produces a magnetic field ?
4. State Ampere's circuital law.
5. What are the advantages of Ampere's law over Biot-Savart's law ?
6. Give importance of Ampere's law.
7. What is solenoid ? Where is it used ?
8. What is toroid ? State two applications of it.
9. Define magnetic flux.
10. State Gauss's law for magnetism.
11. Using formula $\vec{F} = q\vec{v} \times \vec{B}$ and $B = \dfrac{\mu_o i}{2\pi r}$, show that the S.I. units of the magnetic field B and permeability constant may be written as N/A·m and N/A^2 respectively.
12. Magnetic field lines show the direction at every point which a small magnetized needle takes up at that point. Do the magnetic field lines also represent the lines of force on moving charged particle at every point ?   (**Ans.** No, Magnetic force is always normal to $\vec{B}$ $(F = q\vec{v} \times \vec{B})$. It is misleading to call field lines of $\vec{B}$ as lines of force.)
13. Magnetic field lines can be entirely confined within the core of a toroid, but not within solenoid, why ?

    (**Ans.** If field lines were entirely confined between two ends of straight solenoid, the flux through cross-section at each end would be non-zero. But flux of field $\vec{B}$ through any closed surface must always be zero. For toroid, this difficulty does not occur because it has no end.)

(C) Long Answer Type Questions :

1. Define magnetic field and hence explain its properties.
2. Using Biot-Savart's law, obtain expression for magnetic field produced in a long straight conductor.
3. Obtain an expression for $\vec{B}$ on the axis of a current carrying circular coil.
4. State and prove Ampere's circuital law.
5. Obtain an expression for the magnetic field on the axis of a solenoid.
6. What is toroid ? Derive an expression for the magnetic field at a point inside the winding of toroid.
7. What is magnetic flux ? Hence explain Gauss's law for magnetism.
8. Explain the term : Magnetic field lines. Draw field lines produced by permanent magnet, C shaped magnet and straight conductor carrying wire.

(D) Unsolved Problems :

1. A beam of proton ($q = 1.6 \times 10^{-19}$ C) moves at 3×10^5 m/s through a uniform magnetic field with magnitude 4 T, that is directed along Z-axis. The velocity of each proton lies in xy plane at an angle 30° to the Z-axis. Find the force on a proton.   (**Ans.** 9.6×10^{-14} N)
2. A closely wound flat circular coil of 50 turns of wire has a diameter 10 cm and carries a current of 4 ampere. Determine the magnetic induction at the centre of a coil.   (**Ans.** 2.5×10^{-3} tesla)
3. Calculate the magnetic induction in air at a point 10 cm from a long straight wire carrying a current of 5 amperes.   (**Ans.** 1×10^{-5} tesla)
4. A solenoid of length 0.5 m and radius 1 cm is closely wound with 500 turns. Calculate the strength of $\vec{B}$ at a point inside the solenoid if a current of 5 A flows through the winding.

   (Given : $\mu_o = 4\pi \times 10^{-7}$ Wb/A-m)   (**Ans.** 6.28×10^{-3} T)
5. A toroid axis has a circumference of 0.40 m and has total 300 turns of wire. Determine the strength of $\vec{B}$ at points along its axis if a current of 1.5 A flows through the windings.   (**Ans.** 1.414×10^{-3} Wb/m^2)
6. A long, straight conductor carries a current of 1.0 A. At what distance from axis of conductor is the magnetic field caused by current is equal to earth's magnetic field of magnitude 0.5×10^{-4} T ?

      (**Ans.** 4 mm)

7. A straight long vertical wire carries a current of 30 ampere directed upward. Locate a point at which the resultant magnetic field is zero if the horizontal component of earth's field is 2×10^{-5} Wb/m^2.

   (**Hint :** According to right hand screw rule, the field produced by the vertical wire carrying a current is in horizontal plane. At some point (neutral point), this field is balanced by horizontal component of earth's magnetic field. Thus B, I are given to calculate R.) (**Ans.** 0.3 m)

8. A solenoid is 2.0 m long and 3 cm in mean diameter. It has five layers of windings of 1700 turns each and carries a current of 5 amp. Find the value of magnetic induction at its centre.

   $$\left(\textbf{Hint :} \text{ Number of turns per unit length} = \frac{5 \times 1700}{2}\right)$$

   (**Ans.** 2.7×10^{-2} Wb/m^2)

9. A solenoid of length 60 cm is wound uniformly with 6,000 turns of insulated copper wire. If the current passing through the wire is 10 amp., calculate (i) magnetic induction $\overrightarrow{B}$ on the axis of solenoid at the centre, (ii) $\overrightarrow{B}$ on the axis at an end.

   (**Ans.** 0.125 Wb/m^2 and 0.062 Wb/m^2)

10. In a hydrogen atom, a single electron revolves with a frequency 6.6×10^{15} Hz in an orbit of radius 5.3×10^{-11} m. Calculate the magnetic field at the nucleus of hydrogen atom.

    (Given : Charge on electron is 1.6×10^{-19} C). (**Ans.** 2.65 T)

11. A particle with a charge of -1.24×10^{-8} C is moving with instantaneous velocity $\overrightarrow{v} = (4.19 \times 10^4$ m/s$)\, \hat{i} + (-3.85 \times 10^4$ m/s$)\, \hat{j}$. What is the force exerted on this particle by a magnetic field $(1.40$ T$)\, \hat{i}$? (**Ans.** $(-6.68 \times 10^{-4}$ N$)\, \hat{k}$)

12. A circular coil of 20 turns and radius 10 cm is placed in a uniform magnetic field of 0.10 T normal to the plane of the coil. If the current in the coil is 5.0 A, what is the

    (a) total charge on the coil ? (b) total force on the coil ?

    (c) average force on each electron in the coil due to the magnetic field ?

    (The coil is made of copper wire of cross sectional area 10^{-5} m^2, and the free electron density in copper wire is 10^{29} m^{-3})

    (**Ans.** (a) 0, (b) 0, (c) Force $evB = BI/nA = 5 \times 10^{-25}$ N)

❑❑❑

Chapter 5...

Magnetic Properties of Materials

Contents ...

Louis Néel was a French physicist born in Lyon, France. He obtained the degree of Doctor of Science at the University of Strasbourg. He was recipient of the Nobel Prize for Physics in 1970 for his pioneering studies of the magnetic properties of solids. The term ferrimagnetism was coined by Néel, who first studied ferrites systematically on the atomic level.

Louis Néel (1904 – 2000)

His contributions to magnetic materials have found numerous useful applications, particularly in the development of improved computer memory units. Néel had also explained the reasons behind the weak magnetism of certain rocks which made possible the study of the history of Earth's magnetic field.

Introduction

- When a material is placed in a magnetic field it can respond in various ways. On the basis of response, we can classify materials as diamagnetic, paramagnetic, ferromagnetic and anti-ferromagnetic. Properties of these materials are already discussed in third chapter. Most materials are generally considered to be non-magnetic and they become magnetized only in the presence of an applied magnetic field. For example, diamagnetic and paramagnetic materials. In these materials the effects are very weak, and the magnetization is lost, as soon as the external field is removed. When ferromagnetic materials placed in an external magnetic field which not only have a large magnetization, but also retain it even after the removal of the external field. These properties of materials are based on magnetic dipole moment of each atom and their orientations in the material. The effect of the external magnetic field on these dipole moments are studied using different quantities like magnetization, magnetic susceptibility and magnetic permeability.

- On the basis of magnetic properties of materials they have different uses in real life world. Magnets are used in refrigerator doors, VCR, audio cassettes, credit cards, electronic speakers, and microphones, audio head-sets, toys, MRI in hospitals, magnetic storage devices like hard disks and even the inks in the paper money. Some breakfast cereals that are "iron fortified" contain small bits of magnetic materials. One can collect these magnetic materials from slurry of cereals and water with a magnet.

- This chapter provides a brief description of the various magnetic field vectors and magnetic properties and parameters.

5.1 Magnetic Properties

5.1.1 Magnetization ($\vec{M}$)

- Matter is made of atoms. Atoms are made of nuclei and electrons. The electrons are revolving around the nucleus in closed paths and hence constitute electric current loops. Every current loop has a magnetic dipole moment and hence electron in an atom has a magnetic moment due to it's orbital motion. Each electron has a permanent angular momentum also even it is at rest. Such angular momentum is permanent and called the spin angular momentum.

Corresponding to spin each electron has a permanent magnetic moment. The value of permanent magnetic moment due to spin is

$$\mu_s = 9.285 \times 10^{-24} \text{ J/T}$$

- The magnetic moment of nucleus is very smaller than the magnetic moment of an electron. The resultant magnetic moment is the vector sum of all such magnetic moments.
- Generally, the magnetic moments of atoms are randomly oriented. Hence there is no net magnetic moments in any volume of material as shown in Fig. 5.1 (a).

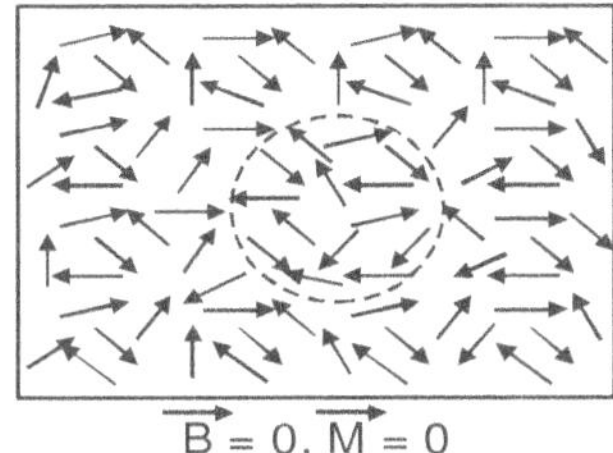

(a) In the absence of external   **(b) In the presence of external**
magnetic field   **magnetic field**

Fig. 5.1

- When the material is kept in external magnetic field, torques act on the atomic dipole. These torques try to align them parallel to the field [Refer Fig. 5.1 (b)]. The alignment along external magnetic field is partial. The thermal motion of atoms frequently changes the orientation of atoms and randomizes the magnetic moments. The alignment of atoms is increased when strength of magnetic field is increased and the temperature is decreased. When the applied field is high, the alignment is perfect and the material is magnetically saturated.
- When atomic dipoles are aligned partially or fully, there is net magnetic moment in the direction of field in any small volume of material.

The magnetic moment per unit volume is called magnetization

$(\overrightarrow{M})$ or intensity of magnetization.

$$\therefore \qquad \overrightarrow{M} = \frac{\text{Magnetic moment}}{\text{Volume}} = \frac{\overrightarrow{\mu}}{V} \qquad \qquad \text{... (5.1)}$$

The unit of magnetic moment is ampere-meter^2 (A·m^2), hence unit of magnetization $\overrightarrow{M}$ is $\dfrac{\text{A·m}^2}{\text{m}^3}$ = A/m.

For bar magnet of pole strength m, length $2l$ and area of cross-section A, the magnetic moment of bar magnet is $\mu = 2m'l$, where m' is the pole strength.

$\therefore$ Intensity of magnetization or simply magnetization is

$$\vec{M} = \frac{\mu}{V} = \frac{2m'l}{A\,(2l)} = \frac{m'}{A} \qquad \text{... (5.2)}$$

From equation (5.2) for a bar magnet, **the intensity of magnetization may be defined as the pole strength per unit area.**

5.1.2 Magnetic Intensity ($\vec{H}$)

- Whenever a magnetic field is applied across a material, it gets magnetized. The actual magnetic field inside it is sum of the magnetic field applied and the intensity due to magnetization.

We define new **vector field $\vec{H}$,** such that

$$\vec{H} = \frac{\vec{B}}{\mu_o} - \vec{M} \qquad \text{... (5.3)}$$

where $\vec{B}$ is the resultant magnetic field and $\vec{M}$ is the intensity of magnetization.

The quantity H given by equation (5.3) is called **magnetic intensity or magnetic field intensity**. The unit of H is same as $\vec{M}$ (i.e. ampere/meter).

If no material is present (in vacuum), $\vec{M} = 0$

$$\vec{H} = \frac{\vec{B}}{\mu_o} \qquad \text{... (5.4)}$$

Whenever the end effects of a magnetized material are neglected, the magnetic field intensity due to magnetization is zero. This may be the case with the ring shaped material or in the middle portion of a long rod. The magnetic intensity is then determined by external sources only, if the material is magnetized.

For example, the magnetic field at the centre of long solenoid having n turns per unit length, carrying current i when no material is kept in it, is

$$\vec{B} = \mu_o \, ni \qquad \qquad \dots (5.5)$$

Hence, magnetic intensity, $\vec{H} = \dfrac{\vec{B}}{\mu_o} = ni \qquad \qquad \dots (5.6)$

When long copper rod is inserted in a long solenoid, the end effects can be neglected. The magnitude of magnetic intensity is same, H = ni.

5.1.3 Magnetic Induction ($\vec{B}$)

- A magnetic induction $\vec{B}$ is defined in terms of $\vec{F_B}$ acting at a point on a test particle with charge q moving through the field with velocity $\vec{v}$.

$$\therefore \qquad \vec{F_B} = q\,\vec{v} \times \vec{B} \qquad \qquad \dots (5.7)$$

Unit of $\vec{B}$ is tesla and denoted by T. $\therefore$ 1 T = 1 N/A·m = 10^4 gauss.

- Magnetic field can be created by number of ways. All of these ways are based on three elementary ways to create B. These are :

  1. Electrical currents (moving charges).

  2. Magnetic dipoles.

  3. Changing electrical field.

- All moving charges produce magnetic field. The magnetic field is generated by a steady current (continuous flow of charges). For example, $\vec{B}$ through a wire is described by the Biot-Savart's law. This is the consequence of Ampere's law. The magnetic field due to permanent magnet is well known. The magnetic field in this case is due to magnetic dipole. The third known source of magnetic field is changing electric field. Just as a changing magnetic field generates electric field, so does a changing electric field generates a magnetic field.

- We have discussed how to calculate magnetic induction ($\vec{B}$) at a point due to solenoid, toroid, long straight conductor and current carrying circular loop in last chapter.

$\vec{B}$ and $\vec{H}$ are related by the following equation :

$$\vec{B} = \mu_o \vec{H} + \mu_o \vec{M} = \mu_o (\vec{H} + \vec{M}) \qquad \ldots (5.8)$$

In vacuum, $\qquad \vec{M} = 0$

$$\vec{B} = \mu_o \vec{H}$$

5.1.4 Magnetic Susceptibility (χ_m)

- For most magnetic materials (i.e. para and diamagnetic materials), it is found that the magnetization $\vec{M}$ is always proportional to the magnetic field intensity $\vec{H}$ i.e., $\vec{M} \propto \vec{H}$

or $\qquad \vec{M} = \chi_m \vec{H} \qquad \ldots (5.9)$

where the constant χ_m **is called as the magnetic susceptibility.**

From equation (5.9), $\chi_m = \dfrac{|\vec{M}|}{|\vec{H}|} \qquad \ldots (5.10)$

Magnetic susceptibility is defined as the ratio of magnetization $\vec{M}$ to the magnetic field intensity $\vec{H}$.

- As M and H have same dimension, the susceptibility χ_m is dimensionless constant. χ_m has no unit.

Using χ_m, we can classify magnetic materials as follows :

For free space or vacuum, $\chi_m = 0$, then material will not respond to any magnetization.

For paramagnetic materials, χ_m is +ve and less than 1.

For diamagnetic materials, χ_m is –ve and less than 1.

For ferromagnetic materials, χ_m is +ve and greater than one, but not constant.

The magnetization M can be written in terms of χ_m and B as

$$\vec{M} = \frac{\chi_m \vec{B}}{\mu_o (1 + \chi_m)} \qquad \ldots (5.11)$$

If χ_m is very small as compared to 1, $\vec{M} \approx \dfrac{\chi_m \vec{B}}{\mu_o}$

Table 5.1 lists magnetic susceptibility for some elements.

Table 5.1 : Magnetic Susceptibility of some elements at room temperature (i.e. 300 K)

Diamagnetic substance	χ_m	Paramagnetic substance	χ_m
Bismuth	-16.6×10^{-5}	Vacuum	0
Copper	-9.8×10^{-6}	Air	0.04×10^{-5}
Carbon (graphite)	-9.9×10^{-5}	Aluminium	2.3×10^{-5}
Diamond	-2.2×10^{-5}	Calcium	1.9×10^{-5}
Gold	-3.6×10^{-5}	Chromium	2.7×10^{-4}
Lead	-1.7×10^{-5}	Lithium	2.1×10^{-5}
Mercury	-2.9×10^{-5}	Magnesium	1.2×10^{-5}
Hydrogen (1 atm)	-9.9×10^{-4}	Niobium	2.6×10^{-5}
Nitrogen (1 atm)	-5.0×10^{-4}	Oxygen (STP)	2.1×10^{-6}
Silver	-2.6×10^{-5}	Platinum	2.9×10^{-4}
Silicon	-4.2×10^{-6}	Tungsten	6.8×10^{-5}

5.1.5 Magnetic Permeability (μ)

The relation between $\vec{B}$, $\vec{H}$ and $\vec{M}$ is

$$\vec{B} = \mu_o (\vec{H} + \vec{M})$$

We know that, $\vec{M} = \chi_m \vec{H}$

$$\therefore \quad \vec{B} = \mu_o (\vec{H} + \chi_m \vec{H}) = \mu_o \vec{H} (1 + \chi_m) \qquad \ldots (5.12)$$

$(1 + \chi_m)$ is constant and is called as relative permeability. It is denoted by μ_r.

$$\mu_r = 1 + \chi_m$$

$$\chi_m = \mu_r - 1 \qquad \ldots (5.13)$$

Relative permeability and $\vec{B}$ are related by

$$\vec{B} = \mu_o \mu_r \vec{H}$$

As μ_o and μ_r are constant, so let $\mu_o \mu_r = \mu$ $\qquad \ldots (5.14)$

μ is called as magnetic permeability of the material.

$$\therefore \quad \vec{B} = \mu \vec{H} \quad \therefore \mu = \frac{\vec{B}}{\vec{H}} \qquad \ldots (5.15)$$

Magnetic permeability (μ) **may be defined as the ratio of magnetic induction** $(\vec{B})$ **to the magnetic intensity** $(\vec{H})$.

Unit of μ **:** From equation $\mu = \dfrac{B}{H}$

$$\text{Unit of } \mu \;=\; \frac{\text{Unit of B}}{\text{Unit of H}} = \frac{\text{weber/meter}^2}{\text{ampere/meter}} = \frac{\text{weber}}{\text{ampere meter}}$$

Since $\qquad$ 1 henry $= \dfrac{1 \text{ weber}}{1 \text{ ampere}}$

$\therefore \qquad$ Unit of $\mu \;=\; \dfrac{\text{henry}}{\text{meter}}$

Thus, unit of μ is $\dfrac{\text{weber}}{\text{amperes} \cdot \text{meter}}$ or $\dfrac{\text{henry}}{\text{meter}}$

But Wb $= $ T m^2 $\therefore$ Other unit of μ is $\dfrac{\text{T m}^2}{\text{A m}} = \dfrac{\text{T m}}{\text{A}}$

Table 5.2 : Classification of materials on the basis of susceptibility and permeability

Material	Susceptibility	Permeability	
		μ_r	μ
Diamagnetic	$-1 \le \chi_m < 0$	$0 < \mu_r < 1$	$\mu < \mu_o$
Paramagnetic	$0 < \chi_m < \varepsilon$	$1 < \mu_r < 1 + \varepsilon$	$\mu > \mu_o$
Ferromagnetic	$\chi_m >> 1$	$\mu_r >> 1$	$\mu >> \mu_o$

Here ε is a small positive number introduced to quantify paramagnetic materials.

5.2 Relation Between $\vec{B}$, $\vec{H}$ and $\vec{M}$

- Consider a toroidal coil (Rowland ring) of N_o turns and carrying a current I_o. Let r_o be the radius of the toroidal coil. To apply Ampere's circuital law, consider a closed path (shown in Fig. 5.2 by a dotted line) inside the toroidal coil (without core material).

$$\therefore \qquad \oint \vec{B} \cdot d\vec{l} \;=\; \mu_o I$$

where I is the current crossing the area enclosed by chosen loop. Since current I_o threads N_o times the closed path, $I = I_o N_o$

$$\therefore \qquad \oint \vec{B} \cdot d\vec{l} \;=\; \mu_o N_o I_o \qquad\qquad \text{... (5.16)}$$

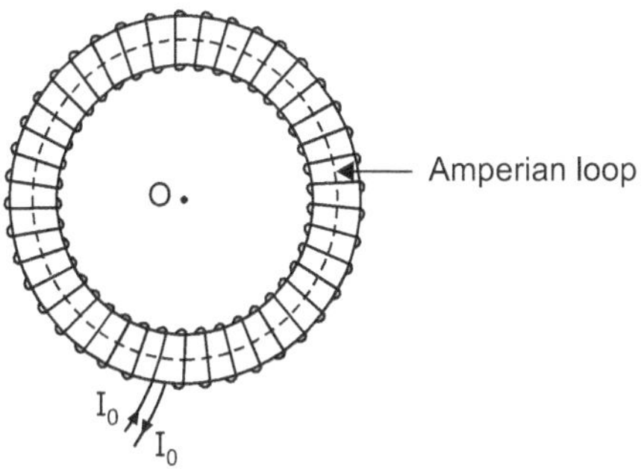

Fig. 5.2 : Toroid

- Let us now put a paramagnetic substance in the volume of toroidal coil. Due to the magnetic field (produced by current I_o) inside the toroidal coil, all the atomic current loops inside the material get oriented along the direction and is perpendicular to the direction of external magnetic field $\vec{B}$. Fig. 5.3 shows the cross-section of magnetized ring consisting of small circular current loops. At any point inside the material, net current is zero because the currents from the adjacent loops cancel each other. However, there is a net current along the surface as there is no cancellation of currents.

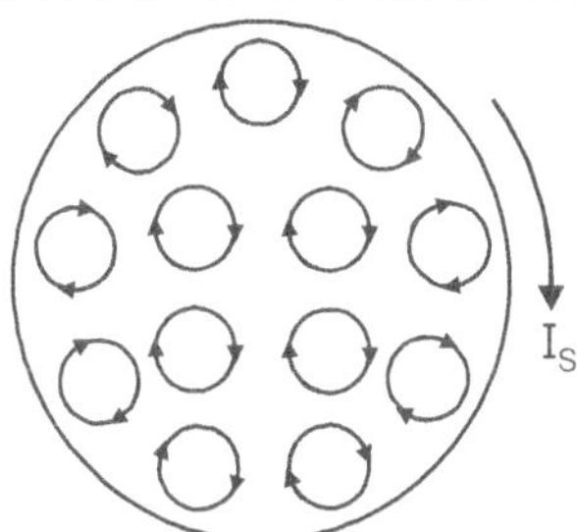

Fig. 5.3 : Cross-section of a magnetized ring

- Thus, the entire network of atomic current is equivalent to a surface current I_S circulating around the surface of the ring. Such a current is known as Amperian surface current. So the magnetic induction at any point on the loop inside the coil in the presence of material is the resultant of magnetic induction due to real current (I_o) and magnetic induction due to surface current (I_S). Thus, the magnetic induction inside the coil in the presence of paramagnetic[*] material is more than without it.

---

[*]   *In case of paramagnetic materials, the direction of I_o and I_S are same.*

- Let for a given current I_o, magnetic induction inside the coil without a material be B and in the presence of core material $B + \Delta B$. The increase in magnetic induction ΔB is due to the surface current.
- The value of B can be increased by ΔB, even in the absence of the core material by increasing the current in the toroid say by an amount I_{mo}. It means that current I_{mo} produces the same magnetic effect (which was produced by surface currents) and therefore surface current can be replaced by effective magnetic current.

The effective magnetic current $I_m = N_o \cdot I_{mo} = (I_{mo} \times$ Number of loops).

Thus, in the presence of magnetic material, Ampere's law may be written as

$$\oint \vec{B} \cdot d\vec{l} = \mu_o (I + I_m) \quad \text{where, } I = N_o I_o$$

$$I_m = N_o I_{mo}$$

$$\therefore \qquad B\, 2\pi r_o = \mu_o N_o I_o + \mu_o N_o I_{mo} \qquad \ldots (5.17)$$

To write above equation in terms of $\vec{M}$, we consider a very small volume element $dV = A\,dl$ of the material.

Magnetization is, $\qquad \vec{M} = \dfrac{\text{Magnetic dipole moment } (d\mu)}{\text{Volume } (dV)}$

Magnetic dipole moment is given by,

$$d\mu = \text{Current} \times \text{Area} \times \text{Number of loops}$$

Number of turns per unit length is $\dfrac{N_o}{2\pi r_o}$.

$$\therefore \qquad d\mu = I_{mo}\, A dl \,\dfrac{N_o}{2\pi r_o} \qquad \ldots (5.18)$$

Using this equation, magnetization $\vec{M}$ can be expressed as

$$\vec{M} = \dfrac{N_o I_{mo}\, A\, dl}{2\pi r_o \cdot A dl} \quad \text{or} \quad N_o I_{mo} = M\, 2\pi r_o \qquad \ldots (5.19)$$

Using equation (5.19) in equation (5.17), we get

$$B\, 2\pi r_o = \mu_o N_o I_o + \mu_o M\, 2\pi r_o$$

$$\text{or} \qquad \oint \vec{B} \cdot d\vec{l} = \mu_o I + \mu_o \oint \vec{M} \cdot d\vec{l}$$

$$\int \dfrac{(\vec{B} - \mu_o \vec{M})}{\mu_o} \cdot d\vec{l} = I \qquad \ldots (5.20)$$

The quantity $\left(\dfrac{\vec{B} - \mu_o \vec{M}}{\mu_o} \right)$ is called **intensity of magnetic field ($\vec{H}$).**

It is also known as magnetizing force.

Hence, in the magnetized material,

$$\frac{\vec{B} - \mu_o \vec{M}}{\mu_o} = \vec{H} \qquad \text{... (5.21)}$$

or $\qquad \vec{B} - \mu_o \vec{M} = \mu_o \vec{H}$

or $\qquad \vec{B} = \mu_o (\vec{H} + \vec{M}) \qquad \text{... (5.22)}$

The equation (5.22) is the relation between $\vec{B}$, $\vec{M}$ and $\vec{H}$.

Using equation (5.21) in equation (5.20), we get

$$\int \vec{H} \cdot d\vec{l} = I \qquad \text{... (5.23)}$$

where $\qquad I = N_o I_o$

From the equation (5.23), the value of H depends only upon the value of real or free current.

The value of magnetization $(\vec{M})$ depends on Amperian surface current which in turn depends on the property of material.

Magnetic field $\vec{B}$ depends on free current as well as surface current.

Thus, Ampere's law in the presence of magnetic material is given by

$$\oint \vec{H} \cdot d\vec{l} = I$$

where I is the real current and does not include the surface current or magnetizing current.

5.3 Hysteresis and Hysteresis Curve

- The magnetization in ferromagnetic material not only depends on magnetic intensity $\vec{H}$, but also history of the specimen. Consider a ring made of ferromagnetic material and placed inside a toroid having n turns per unit length. As current $\vec{I}$ passes through a toroid, the magnetic intensity H is produced in it. The magnetic field produced is $\qquad \vec{B}_o = \mu_o n \vec{I}$

$$\vec{H} = \frac{\vec{B}_o}{\mu_o} = n \vec{I} \qquad \text{... (5.24)}$$

- $\vec{B_o}$ is the field produced by toroid only. The ring gets magnetized and produces extra field due to magnetization.

  The total field in the ring is

  $$\vec{B} = \mu_o (\vec{H} + \vec{M})$$

  $$\vec{M} = \frac{\vec{B}}{\mu_o} - \vec{H} = \frac{\vec{B}}{\mu_o} - n\,\vec{I} \qquad \qquad ...\,(5.25)$$

- One can measure the total field B inside the ring using apparatus called Rowland ring. The intensity of magnetization can be obtained from above equation (5.25). Thus from equations (5.24) and (5.25), one can obtain $\vec{H}$ for every $\vec{M}$ for any current.

- The relation between $\vec{B}$ and $\vec{H}$ in ferromagnetic materials is complex. It is often not linear and it depends on the ferromagnetic specimen. Fig. 5.4 depicts the behaviour of the material as we take it through one cycle of magnetization.

- Let the material (specimen) be unmagnetized initially. We place it in a solenoid and increase the current through the solenoid. The magnetic field $\vec{B}$ in the material rises and saturates (as depicted in the curve Oa). This behaviour represents the alignment and merger of domains. It is pointless to increase the current (and hence the magnetic intensity $\vec{H}$) beyond this saturation.

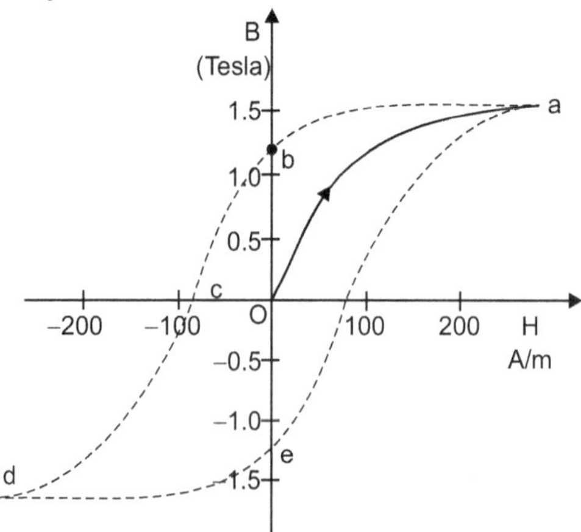

Fig. 5.4 : The magnetic hysteresis loop is the B-H curve for ferromagnetic materials

- Next, we decrease H and reduce it to zero. At point b, H= 0, but $B \neq 0$. This is represented by the curve ab. The value of B at H = 0 is called **retentivity or remanence**.
- In figure, value of $B_R \sim 1.2$ T, where the subscript R denotes retentivity. The domains are not completely randomized even though the external field has been removed.
- Next, the current in the solenoid is reversed and slowly increased. Certain domains are flipped until the net field inside stands nullified. This is represented by the curve bc. The value of H at point c is called **coercivity**.
- In Fig. 5.4, value of $H \sim -90$ Am^{-1}. The value of H needed to make M = 0 is called **coercive force**. In figure the value of coercive force is represented by magnetic field of H corresponding to Oc.
- As the reversed current is increased in magnitude, we once again obtain saturation (curve cd). The saturated magnetic field is $B_s \sim 1.5$ T. Next, the current is reduced (curve de) and reversed (curve ea). The cycle repeats itself.

We make the following observations :

- The curve Oa does not retrace itself though H is reduced. For a given value of H, B is not unique, but depends on the previous history of the sample. This phenomenon is called **hysteresis.** The word hysteresis means *'lagging behind'* (and not 'history'). The curve abcdea is called the **hysteresis loop**. The area of hysteresis loop is proportional to the thermal energy developed per unit volume of material as it goes through hysteresis cycle.
- In Fig. 5.4, when $B \approx 1.5$ T, we obtain $B/\mu_o \approx 1.2 \times 10^6$ Am^{-1}. However, H at this value of B is merely 200 Am^{-1}. This implies that the effective relative magnetic permeability $\mu_r \approx 10000$! Such large values are characteristic of ferromagnetic materials only.
- No segment, Oa, ab, etc. of the curve is linear showing B is not proportional to H over any appreciable range.
- Note that $BH = B^2/(\mu_o \, \mu_r)$ has the dimensions of energy per unit volume. The area within the B-H loop of Fig. 5.4 represents the energy dissipated per unit volume in the material. The 'source' of energy is the chemical battery which drives the current in the solenoid. The 'sink' is the Joule heating (i^2R) in the solenoid wire and the dissipative hysteresis heat loss in the magnetic material.

Hysteresis can be understood on the basis of magnetic domain.

- It has been observed that the domain boundaries and reorientation of domain direction are not totally reversible. When applied field is increased and then decreased back to initial value, the domains do not return completely to their original configuration. But, they retain some memory of their alignment after the initial increase.

- The memory of magnetic material is essential for the magnetic storage of information. There are a great variety of applications of the hysteresis in ferromagnets. Many of these make use of their ability to retain a memory, for example magnetic tapes, hard disks and credit cards. In these applications, *hard* magnets (high coercivity).

- *Soft* magnets (low coercivity) are used as cores in electromagnets. The non-linear response of the magnetic moment to a magnetic field boosts the response of the coil wrapped around it. The low coercivity reduces the energy loss associated with the hysteresis.

- When lightening in the sky sends current along multiple tortuous paths through the ground, the current produces intense magnetic fields that can suddenly magnetize any ferromagnetic material in the nearby rock. Because of hysteresis, such rocks retain magnetism after lightning strikes. When pieces of such rock are exposed, broken and loosen by weather changes. Finally, they become loadstones.

- The size and shape of the hysteresis curve for ferromagnetic and a ferrimagnetic material is of considerable practical importance. The area within a loop represents a magnetic energy loss per unit volume of material per magnetization–demagnetization cycle; this energy loss is manifested as heat that is generated within the magnetic specimen and is capable of raising its temperature.

- Both ferromagnetic and ferrimagnetic materials are classified as either *soft* or *hard* on the basis of their hysteresis characteristics.

- A **hard ferromagnetic** material like steel has a wide hysteresis loop as shown in Fig. 5.5 (a). The area of hysteresis loop is also larger for steel than soft iron. Hard magnetic materials are utilized in permanent magnets, which must have a high resistance to demagnetization. In terms of hysteresis behaviour, it has a high remanence, coercivity, and saturation flux density, as well as a low initial permeability, and high hysteresis energy losses.

- The two most important characteristics relative to applications for these materials are the coercivity and the "energy product," designated as $(BH)_{max}$. This $(BH)_{max}$ corresponds to the area of the largest *B-H* rectangle that can be constructed within the second quadrant of the hysteresis curve. The value of the energy product is representative of the energy required to demagnetize a permanent magnet; that is, the larger the $(BH)_{max}$, harder is the material in terms of its magnetic characteristics.

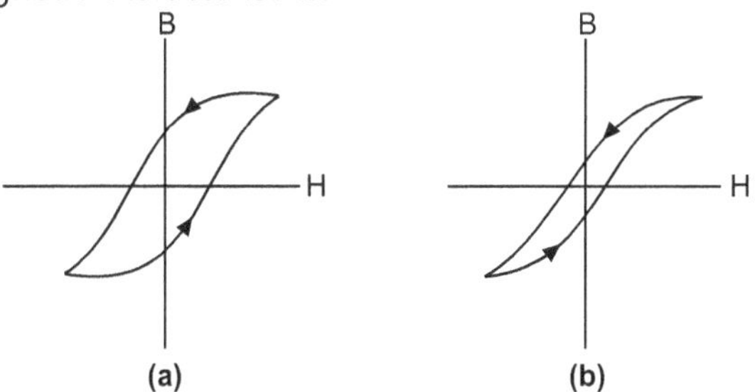

Fig. 5.5 : The magnetic hysteresis loop for (a) hard, (b) soft ferromagnetic material

- A **soft ferromagnet** like soft iron has a narrow hysteresis loop as depicted in Fig. 5.5 (b). The energy loss is minimal and at the same time the value of μ_r enhances the magnetic field.
- The material of soft iron has low retentivity, low coercivity and have a high initial permeability .
- **Soft magnetic materials** are devices that are subjected to alternating magnetic fields and in which energy losses must be low; one familiar example consists of transformer cores. Such 'soft' magnetic materials are useful as electromagnets and also used as cores in solenoids and transformers where field reversals are frequent.
- Another property consideration for soft magnetic materials is electrical resistivity. In addition to the hysteresis energy losses described above, energy losses may result from electrical currents that are induced in a magnetic material by a magnetic field that varies in magnitude and direction with time; these are called *eddy currents*.
- It is most desirable to minimize these energy losses in soft magnetic materials by increasing the electrical resistivity. This is accomplished in ferromagnetic materials by forming solid solution alloys. Iron-silicon and iron–nickel alloys are examples.
- The ceramic ferrites are commonly used for applications requiring soft magnetic materials because they are intrinsically electrical insulators. Their applicability is somewhat limited.

- The hysteresis characteristics of soft magnetic materials may be enhanced for some applications by an appropriate heat treatment in the presence of a magnetic field. Using such a technique, a square hysteresis loop may be produced, which is desirable in some magnetic amplifier and pulse transformer applications.
- In addition, soft magnetic materials are used in generators, motors, dynamos, and switching circuits.

5.4 Ferrite Materials and Their Applications

- The term ferrite is commonly used generically to describe a class of magnetic oxide compounds that contain iron oxide as a principal component. Ferrite materials are usually ferrimagnetic ceramic compounds derived from iron oxides. A ferrite is formed by the reaction of ferric oxide (Fe_2O_3 or rust) with any other metal such as magnesium, aluminium, barium, manganese, copper, nickel, cobalt, or even iron itself. For example, the following reaction at high temperature gives zinc ferrite.

$$Fe_2O_3 + ZnO \rightarrow ZnFe_2O_4$$

- A ferrite is usually described by the formula $M(Fe_xO_y)$, where M represents any metal that forms divalent bonds. Commonly known ferrites are nickel ferrite ($NiFe_2O_4$), manganese ferrite ($MnFe_2O_4$), and magnetite (Fe_3O_4). Magnetite is also known as *loadstone*, which is a genuine ferrite and was the first magnetic material known to the ancient people.
- The ferrites are electrically non-conductive, meaning that they are insulators. They have high magnetic permeability and can easily be magnetized or attracted to a magnet. Ferrites are generally gray or black. They are hard, brittle and polycrystalline *i.e.,* made up of a large number of small crystals.
- Ferrites can be divided into two families based on their magnetic coercivity or resistance to being demagnetized. These types are hard ferrites and soft ferrites. Hard ferrites have high coercivity, so are difficult to demagnetize. Hard ferrites are used to make permanent magnets for applications such as refrigerator magnets, loudspeakers, and small electric motors. Soft ferrites have low coercivity, so they easily change their magnetization and act as conductors of magnetic fields. Soft ferrites are used in the electronics industry to make efficient magnetic cores called ferrite cores for high-frequency inductors and transformers, and in various microwave components.

Applications of Ferrites: Ferrites are regarded as better magnetic materials than pure metals because of their high resistivity, lower cost, easier manufacturing process and superior magnetization properties. **Some of the applications are**

(1) Cores of ferrite materials are used in electronic inductors, transformers and electromagnets because the high electrical resistance of the ferrite leads to very low eddy current losses.

(2) A number of ferrites absorb microwave energy in only one direction or orientation; due to this they are used in microwave wave guides.

(3) They are used for Computer Hard Drive Magnets, Microphones, Headphones. Loudspeakers.

(4) Ferrites are used as magnetic head transducer in recording.

(5) Ferrite powders are used in the coatings of magnetic recording tapes.

Solved Problems

Problem 5.1 : *A bar magnet made of iron has magnetic moment 2.0 A.m^2 and mass 5×10^{-3} kg. If the density of iron is 6×10^{-3} kg/m^3, find the intensity of magnetization.*

Solution : $\text{Volume} = \dfrac{\text{Mass}}{\text{Density}} = \dfrac{5 \times 10^{-3} \text{ kg}}{6 \times 10^{-3} \text{ kg/m}^3} = 0.8333 \text{ m}^3$

$$\text{Intensity of magnetization, } \vec{H} = \frac{\vec{M}}{\vec{V}} = \frac{2 \text{ A.m}^2}{0.8333 \text{ m}^3} = \textbf{2.4 A/m} \quad \textbf{... Ans.}$$

Problem 5.2 : *The susceptibility of tungsten at 27 °C is 6.8×10^{-5}. Find the percent increase in magnetic field B when the space within a current carrying toroid is filled with tungsten.*

Solution : In absence of tungsten, $B_o = \mu_o H$

When space inside the toroid is filled with tungsten,

$$B = \mu H = \mu_o (1 + \chi_m) H$$

$\therefore$ Increase in field is $B - B_o = \mu_o \chi_m H$

The percentage increase in B is

$$\frac{B - B_0}{B_0} \times 100 = \frac{\mu_o \chi_m H}{\mu_o H} \times 100 = \chi_m \times 100$$

$$= 6.8 \times 10^{-5} \times 100 = \textbf{6.8} \times \textbf{10}^{-3} \qquad \textbf{... Ans.}$$

Problem 5.3 : *An ideal solenoid of aluminium core, have 40 turns per cm and a current 2.5 A. Calculate magnetization M developed in the core and the magnetic field at the centre.*

Predict nature of substance. (Given : $\chi_{aluminium} = 2.3 \times 10^{-5}$).

Solution : Given : n = 40 turns/cm = 4000 turns/m

Magnetic intensity, H = nI = 4000 turns/m $\times$ 2.5 A = 10,000 A/m

Magnetization, M = $\chi_m H$ = $2.3 \times 10^{-5} \times 10{,}000$ = **0.23 A/m** ... **Ans.**

Magnetic field, B = $\mu_o (H + \overrightarrow{M})$ = $4\pi \times 10^{-7} (10000 + 0.23)$

$\qquad\qquad$ = $12.56 \times 10^{-7} \times 10000.23$ = **1.256×10^{-2} T** ... **Ans.**

Here, $H >> M$, hence substance is **paramagnetic.**

Problem 5.4 : *The maximum value of the permeability of some metals is 0.150 T-m/A. Find the value of maximum relative permeability and susceptibility.*

Solution : Relative permeability,

$$\mu_r = \frac{\mu}{\mu_o} = \frac{0.150 \text{ T-m/A}}{4\pi \times 10^{-7} \text{ T-m/A}} = 1.19 \times 10^5$$

Susceptibility, χ_m = $\mu_r - 1 = 1.19 \times 10^5 - 1 \approx$ **1.19×10^5** ... **Ans.**

Problem 5.5 : *A solenoid of 500 turns/m is carrying a current of 3 A. If the core is made of iron which has a relative permeability of 5000, determine the magnitude of the magnetic intensity, magnetization and the magnetic field inside the core.*

Solution : Given : n = 500 turns/m, I = 2 A, μ_r = 5000

The magnetic intensity, $H = nI = 500 \times 3$ = **1500 A/m**

Relative permeability, $\mu_r = 1 + \chi_m$

$$\chi_m = \mu_r - 1 = 4999 \approx 5000$$

$$\mu_r = \frac{\mu}{\mu_o}$$

$\therefore\qquad\qquad \mu = \mu_o \cdot \mu_r = 5000\,\mu_o$

$\therefore$ Magnetization is $M = \chi_m H$

$\qquad\qquad$ = 5000×1500 = **7.5×10^6 A/m** $\qquad$ ... **Ans.**

Magnetic field is B = $\mu_o \mu_r H = 4\pi \times 10^{-7} \times 5000 \times 1500$

$\qquad\qquad$ = **9.42 T** $\qquad\qquad\qquad\qquad$ ... **Ans.**

Problem 5.6 : *A tightly-wound, long solenoid having 100 turns/cm carries a current of 3.00 A. (a) Find the magnetic intensity H and magnetic field B at the centre of solenoid. (b) What is the value of H and B if iron core is inserted in the solenoid and magnetization M in the core is 4×10^4 A/m ?*

Solution : (a) Magnetic intensity H at the centre of solenoid,

$\qquad\qquad H$ = $nI = 100 \times 10^2 \text{ m}^{-1} \times 3$ A

$\qquad\qquad H$ = **3×10^4 A/m** $\qquad\qquad\qquad$ ... **Ans.**

Magnetic field, B = $\mu_o H = 4\pi \times 10^{-7} \times 3 \times 10^4 = 12\pi \times 10^{-3}$ T

$\qquad\qquad B$ = **37.68×10^{-3} T** $\qquad\qquad\qquad$ ... **Ans.**

(b) The value of **H does not change** as the iron core is inserted and H remains the same. H $=$ **3×10^4 A/m**   ... **Ans.**

But magnetic field B becomes

$$B = \mu_o (H + M) = (4 \times \pi \times 10^{-7} \text{ T m-A}) (3 \times 10^4 + 4 \times 10^4) \text{ A/m}$$
$$= 4\pi \times 10^{-7} \times 7 \times 10^4 = \textbf{8.79} \times \textbf{10}^{-2} \textbf{ T} \qquad \text{... \textbf{Ans.}}$$

Problem 5.7 : *A domain in ferromagnetic iron is in the form of a cube of side length 1 mm. Estimate the number of iron atoms in the domain and the maximum possible dipole moment and magnetization of the domain. The molecular mass of iron is 55 g/mole and its density is 7.9 g/cm³. Assume that each iron atom has a dipole moment of 9.27×10^{-24} Am².*

Solution : Given : $l = 1$ mm $= 1 \times 10^{-6}$ m, $\rho = 7.9$ g/cm³,
$$\mu = 9.27 \times 10^{-24} \text{ A m}^2.$$

The volume of the cubic domain is
$$V = (1 \times 10^{-6})^3 = 1 \times 10^{-18} \text{ m}^3 = 10^{-12} \text{ cm}^3$$

Its mass is volume $\times$ density.
$$\therefore \qquad m = 7.9 \times 10^{-12} \text{ g}$$

It is given that an Avogadro number (6.023×10^{23}) of iron atoms has a mass of 55 g. Hence, the number of atoms in the domain is

$$N = \frac{7.9 \times 10^{-12} \times 6.023 \times 10^{23}}{55}$$

$$N = \textbf{8.65} \times \textbf{10}^{10} \textbf{ atoms} \qquad \text{... \textbf{Ans.}}$$

The maximum possible dipole moment μ_{max} is achieved for the (unrealistic) case when all the atomic moments are perfectly aligned. Thus,

$$\mu_{max} = N\mu$$
$$= (8.65 \times 10^{10}) \times (9.27 \times 10^{-24})$$
$$= 8.02 \times 10^{-13} \text{ Am}^2$$
$$\mu_{max} \approx \textbf{8} \times \textbf{10}^{-13} \textbf{ Am}^2 \qquad \text{... \textbf{Ans.}}$$

The consequent magnetization is

$$M_{max} = \frac{\mu_{max}}{\text{Domain volume}} = \frac{8 \times 10^{-13}}{1 \times 10^{-18}}$$

$$M_{max} = \textbf{8} \times \textbf{10}^5 \textbf{ Am}^{-1} \qquad \text{... \textbf{Ans.}}$$

Problem 5.8 : *A solenoid of 400 turns/m is carrying a current of 2 A. Its core is made of iron which has a relative permeability of 5000. Determine the magnitudes of the magnetic intensity, magnetization and the magnetic field inside the core.*

Solution : The magnetic intensity is,

$$H = nI$$
$$H = 400 \text{ m}^{-1} \times 2 \text{ A} = \textbf{800 A m}^{-1} \qquad \text{... \textbf{Ans.}}$$

It is given that $\mu_r = 5000$.

$$\mu_r = 1 + \chi_m, \text{ i.e., } \chi_m = 4999 \approx 5000$$

$$\mu = \mu_o \mu_r = 5000 \, \mu_o$$

Hence, the magnetization is,

$$M = \chi_m H = 5000 \times 800 = \mathbf{4 \times 10^6 \ A \ m^{-1}} \qquad \text{... Ans.}$$

The magnetic field is,

$$B = \mu_o \mu_r H = \mu H = 5000 \, \mu_o \, H$$

$$B = 5000 \times 4\pi \times 10^{-7} \times 800 = \mathbf{5.024 \ T} \qquad \text{... Ans.}$$

Problem 5.9 : *A sample of paramagnetic salt contains 2.0×10^{24} atomic dipoles each of dipole moment 1.5×10^{-23} J/ T. The sample is placed under homogeneous magnetic field of 0.64 T and cooled to temperature 4.2 K. The degree of magnetic saturation achieved is 15%. What is the total dipole moment of the sample for a magnetic field of 0.98 T and temperature 2.8 K (assuming Curie law) ?*

Solution : Given : $\mu = 1.5 \times 10^{-23}$ J/T, Number of dipoles $= 2 \times 10^{24}$.

Initially total magnetic moment $= 0.15 \times 1.5 \times 10^{-23} \times 2.0 \times 10^{24}$
$= 4.5$ J/T

Using Curie law, $\vec{M} = C \dfrac{\vec{B}}{T}$

$$M \propto \dfrac{B}{T}$$

Final dipole moment $= 4.5 \times \left(\dfrac{0.98}{0.64}\right)\left(\dfrac{4.2}{2.8}\right) = \mathbf{10.34 \ J/T} \qquad \text{... Ans.}$

Think Over It

1. How can you identify the material is soft magnetic or hard magnetic material on the basis of hysteresis curve ?
2. What exactly causes magnetic properties of materials ?

Summary

- Magnetization is $\vec{M} = \dfrac{\vec{\mu}}{V}$ and $\vec{M} = \chi_m \vec{H}$

where χ_m is magnetic susceptibility and $\vec{H}$ is magnetic intensity

- Magnetic intensity $(\vec{H})$, $\vec{H} = \dfrac{\vec{B_o}}{\mu_o} = \dfrac{\mu_o\, ni}{\mu_o} = ni$

- Magnetic field inside the solenoid, $B = \mu_o ni$

$$\vec{B} = \mu_o\,(1 + \chi_m)\,\vec{H} = \mu_o\,\mu_r\,\vec{H} = \mu\vec{H}$$

where $\quad \mu$ = relative magnetic permeability

$\qquad\qquad \mu_r = \mu_o\,(1 + \chi_m)$

- The relation between $\vec{B}$, $\vec{H}$ and $\vec{M}$ is

$$\vec{B} = \mu_o\,(\vec{H} + \vec{M}) = \mu_o\,\vec{H}\,(1 + \chi_m)$$

- Magnetic susceptibility, $\chi_m = \dfrac{|\vec{M}|}{|\vec{H}|}$

- Magnetic permeability, $\mu = \dfrac{B}{H}$

- Magnetic materials are broadly classified as : diamagnetic, paramagnetic, and ferromagnetic. For diamagnetic materials, χ_m is negative and generally very small, for paramagnetic, χ_m is positive and small. Ferromagnetic materials have large χ_m characterized by non-linear relation between $\vec{B}$ and $\vec{H}$.

- Lagging behind of $\vec{B}$ behind $\vec{H}$ is called hysteresis. Area of hysteresis loop represents the energy dissipated per unit volume in the material.

- Hard ferromagnetic material has wide hysteresis loop while a soft ferromagnetic material has narrow hysteresis loop.

- The substances, which at room temperature, retain their ferromagnetic property for long periods of time are called permanent magnets.

Exercises

(A) Multiple Choice Questions :

1. The ratio of intensity of magnetization to the magnetization force is known as ......

   (a) flux density   (b) susceptibility

   (c) relative permeability   (d) magnetic moment

2. The unit of relative permeability is ......
    (a) henry/metre   (b) henry-metre
    (c) henry/sq.m   (d) it is dimensionless
3. Ferrites are magnetically ......
    (a) non-magnetic materials   (b) ferro-magnetic materials
    (c) paramagnetic materials   (d) ferri-magnetic materials
4. For which of one the following materials the saturation value is the highest ?
    (a) ferromagnetic   (b) paramagnetic
    (c) diamagnetic   (d) ferrites
5. The magnetic susceptibility is negative for ......
    (a) ferromagnetic material only
    (b) paramagnetic and ferromagnetic materials
    (c) diamagnetic material only
    (d) paramagnetic material only
6. The magnetic susceptibility of a paramagnetic material is ......
    (a) less than zero   (b) less than one but positive
    (c) greater than one   (d) zero
7. Which of the following is true for soft magnetic materials ?
    (a) high permeability and low coercive force
    (b) low permeability and high coercive force
    (c) high residual magnetism
    (d) low resistivity
8. For which of the following materials the net magnetic moment should be zero ?
    (a) diamagnetic   (b) ferrimagnetic
    (c) antiferromagnetic   (d) antiferrimagnetic

Answers

1. (b)	2. (d)	3. (d)	4. (a)	5. (c)	6. (b)	7. (a)	8. (a)

(B) Short Answer Type Questions :
1. What is Bohr magneton ?
2. Define : (a) Magnetization, (b) Magnetic intensity, (c) Magnetic induction.
3. Define the terms : Magnetic permeability and Susceptibility.
4. Explain magnetic susceptibility.
5. Write the relation between $\vec{B}$, $\vec{M}$ and $\vec{H}$.
6. Define remanence and coercivity.
7. Define hysteresis and draw hysteresis curve.
8. What are ferrite materials ?

(C) Long Answer Type Questions :

1. Explaining how material is magnetized, obtain an expression for Bohr magneton.

2. Explain in detail : (a) Magnetization ($\vec{M}$), (b) Magnetic intensity ($\vec{H}$), (c) Magnetic field ($\vec{B}$).

3. Explain the terms : Magnetic susceptibility and Magnetic permeability.

4. Obtain the relation between $\vec{B}$, $\vec{M}$ and $\vec{H}$. Discuss qualitatively.

5. What is Curie temperature ? Explain the relation between Curie temperature and susceptibility.

6. Distinguish between paramagnetic, diamagnetic and ferromagnetic materials on the basis of susceptibility and permeability.

7. What is hysteresis ? Using hysteresis curve explain the terms retentivity and coercivity.

8. Briefly describe the phenomenon of magnetic hysteresis, and why it occurs for ferromagnetic and ferrimagnetic materials.

9. Explain hard ferromagnet and soft ferromagnet using hysteresis.

10. Cite the differences between hard and soft magnetic materials in terms of both hysteresis behaviour and typical applications.

11. What are ferrite materials ? Give their applications.

(D) Unsolved Problems :

1. A long cylindrical iron core of cross-sectional area 5.00 cm^2 is inserted into long solenoid having 2000 turns/m and carrying a current 2.00 A. The magnetic field inside the core is found to be 1.57 T. Neglecting end effects, find magnetization M of the core and the pole strength developed.   (**Ans.** 1.25×10^6 A/m, 625 A.m)

2. The susceptibility of magnesium at 300 K is 1.2×10^{-5}. At what temperature will the susceptibility increase to 1.8×10^{-5} ? (**Ans.** 200 K)

3. The magnetic field $\vec{B}$ and magnetic intensity $\vec{H}$ in a material are found 1.6 T and 1000 A/m respectively. Calculate relative permeability μ_r and susceptibility χ_m of material.   (**Ans.** 1.3×10^3 each)

4. A toroid has a mean radius R equal to $20/\pi$ cm and 400 turns of wire carrying a current of 2.0 A. An aluminium ring at temperature 280 K inside the toroid provides the core.

(a) If magnetization $\vec{M}$ is 4.8×10^{-2} A/m, find susceptibility of aluminium at 280 K.

(b) If the temperature of the aluminium ring is raised to 320 K, what will be the magnetization ?

$$\left(\textbf{Hint : } n = \frac{400}{2\pi R}, \ H = nI, \ \chi_m = \frac{M}{H}, \ \frac{\chi_2}{\chi_1} = \frac{T_1}{T_2}\right)$$

(**Ans.** (a) 2.4×10^{-5}, (b) $M = 4.2 \times 10^{-2}$ A/m)

5. The maximum value of the permeability of μ metal (77% Ni, 16% Fe, 5% Cu, 2% Cr) is 0.126 T-m/A. Find the maximum relative permeability and susceptibility.   (**Ans.** 1×10^5 each)

6. A permanent magnet is made of ferromagnetic material with a magnetization M of about 8×10^5 A/m. The magnet is in the shape of a cube of side 2 cm. Find the magnetic dipole moment of the magnet.   (**Ans.** 6.4 A.m^2)

7. A Rowland ring of mean radius 15 cm has 3500 turns of wire wound on a ferromagnetic core of relative permeability 800. What is the magnetic field B in the core for a magnetizing current of 1.2 A ?

[**Hint :** Use $B = \mu_o K \, ni$, $K = \mu/\mu_o$ = relative permeability]

(**Ans.** $B = 4.48$ T)

8. A solenoid has a core of material with relative permeability 400. The windings of solenoid are insulated from core and carry a current of 2 A. If number of turns are 1000 per metre, calculate H, M and B.

(**Ans.** $H = 2 \times 10^3$ A/m, $B = 1$ T, $M = 8 \times 10^5$ A/m)

9. Nitric oxide (NO) is a paramagnetic compound. The magnetic moment of each NO molecule has a maximum component in any direction of about one Bohr magneton. Compare the interaction energy of such magnetic moments in a 1.5 T magnetic field with the average translational K.E. of molecule at 300 K.

[**Hint :** $U = -U_B B = 1.4 \times 10^{-23}$ J, Av. K.E. $= \frac{3}{2} kT = 6.2 \times 10^{-21}$ J]

10. A solenoid of 500 turns/m is carrying a current of 3 A. Its core is made of iron which has a relative permeability of 5000. Determine magnitudes of the magnetic intensity, magnetization and magnetic field inside the core.

($H = nI = 1500$ Am^{-1}, $M = \chi H = 7.5 \times 10^6$ Am^{-1},

$B = \mu H = 5000 \mu_0 H = 9.4$ T)

❏❏❏

MODEL QUESTION PAPERS

MODEL QUESTION PAPER-I

Total Marks : 35   **Duration : 2 Hours**

Note : (a) Q.1 is compulsory.

(b) Solve any three questions from Q.2 to Q.5.

(c) Questions 2 to 5 carry equal marks.

1. **Solve any Five of the following :**   **(5)**

    (a) State superposition principle in electrostatics.

    (b) What is meant by polar molecule? Give example.

    (c) What is the effect of magnetic field on dielectric material ?

    (d) State Ampere's circuital law.

    (e) Two charges 1 μC and 2 μC are separated by 0.01 m. What is the force of repulsion between them ?

    (f) The maximum value of the permeability of some metals is 0.150 T-m/A. Find the value of maximum relative permeability.

2. (A) (i) Using Biot-Savart's law, obtain magnetic induction due to a long straight conductor.   **(6)**

        (ii) State the difference between diamagnetic and paramagnetic materials.

OR

        (iii) Explain the concept of electric field.

    (B) (i) Define magnetization and magnetic intensity.   **(4)**

        (ii) Explain dielectrics of non-polar molecules.

3. (A) (i) State and prove Gauss's law in electrostatics.   **(6)**

        (ii) What is meant by electric polarization of dielectric material ? Define polarization vector.

OR

        (iii) Explain anti-ferromagnetic materials.

    (B) (i) Define electric dipole and dipole moment.   **(4)**

        (ii) Calculate the electric intensity due to point charge 2×10^{-10} C at a distance 1 m from the charge.

4. (A) (i) Using Ampere's circuital law, obtain magnetic induction due to a long solenoid.   **(6)**

(P.1)

(ii) Write a note on magnetic material.

OR

(iii) Explain the concept of electric flux.

(B) (i) Find the energy due to system of three charges.   (4)

(ii) A coil of 20 cm radius has 15 turns and carries a current of 3 ampere. Find the magnetic field at the centre of the coil.

(Given : $\mu_o = 4\pi \times 10^{-7}$ Wb A^{-1} m^{-1})

5. Write short notes on any Four of the following :   (10)

(A) Discuss variation of force between two charges with distance.

(B) What are ferrite materials ? Give two applications.

(C) What is electric displacement vector ? What is its relation with electric intensity ?

(D) Show that the torque is acting on the dipole when placed in electric field.

(E) What are soft and hard magnetic materials ?

(F) Explain the magnetic field.

❏❏❏

MODEL QUESTION PAPER-II

Total Marks : 35   **Duration : 2 Hours**

1. Solve any Five of the following :   (5)

(a) State Coulomb's law in electrostatics.

(b) What is meant by non-polar molecule? Give example.

(c) What is the effect of magnetic field on moving charged particle ?

(d) State Gauss's law in dielectrics.

(e) Find the electric intensity due to 1 μC charge at a distance 0.01 m from the charge.

(f) A solenoid of 500 turns/m is carrying a current of 2 A. If the core is made of iron which has a relative permeability of 5000, determine the magnitude of the magnetic intensity.

2. (A) (i) Using Biot-Savart's law, obtain magnetic induction due to circular coil.   **(6)**

(ii) State the comparison between ferromagnetic and antiferromagnetic materials.

OR

     (iii) Explain the superposition principle.

(B) (i) Explain paramagnetic material. **(4)**

     (ii) Explain effect of electric field on dielectric material.

3. (A) (i) Obtain equation of potential at a point due to electric dipole. **(6)**

     (ii) Explain magnetic susceptibility.

OR

     (iii) Explain Bohr magneton.

(B) (i) Explain electric dipole moment.

     (ii) An electric dipole consisting of two opposite charges each of magnitude 2.00 μC is separated by a distance of 2.0 cm. The dipole is placed in an external field of intensity 1.0×10^5 N/C. Calculate the maximum torque on the dipole..

4. (A) (i) Using Ampere's circuital law, obtain magnetic induction due to a long straight wire. **(6)**

     (ii) Write a note on ferrites.

OR

     (iii) Explain the concept of magnetic field.

(B) (i) Find the electric potential at a point due to point charge. **(4)**

     (ii) A charge of 9×10^{-9} C is placed inside a cube. Calculate the electric flux linked with the cube.

5. Write short notes on any Four of the following : **(10)**

(A) Discuss energy of system of charges.

(B) Give the applications of ferrite materials.

(C) What is electric polarization vector ? What is its unit ?

(D) Define electric displacement vector $\vec{D}$. Give its SI unit.

(E) Define the terms : Magnetic intensity and Magnetic permeability.

(F) Explain the hysteresis curve.

❑❑❑